BY THE SAME AUTHOR / TINCRAFT FOR CHRISTMAS

TINCRAFT

by Lucy Sargent

DESIGNS AND DRAWINGS BY THE AUTHOR
FLOWER ARRANGEMENTS BY MARIANNA BROCKWAY
PHOTOGRAPHS BY GEORGE C. BRADBURY

Simon and Schuster • New York

Published by Simon and Schuster
Rockefeller Center, 630 Fifth Avenue
New York, New York 10020

First printing

SBN 671-21225-7
Library of Congress Catalog Card Number: 72-83921
A Helen Van Pelt Wilson Book
Designed by Betty Crumley
Manufactured in the United States of America
Printed by The Murray Printing Company, Forge Village, Mass.
Bound by H. Wolff, New York

A Nosegay for You, Marianna

HOW IT ALL BEGAN

Wandering through her fern garden one Sunday after church, Helen Van Pelt Wilson, my friend and editor, said casually, "Somebody should write a book about things you can make out of tin cans." I agreed, "Somebody should." And that put me in mind of an article from Woman's Day *on Christmas ornaments made from tin which was moldering away in my files at that very moment beside a bushel basket of rusting cans. But that very moment came and went, and our thoughts and footsteps wandered to other subjects and other places.*

A few weeks later another friend and author said casually, "You know, I'm editing a children's series of folk tales from around the world. How would you like to do the stories on Thailand?"

That seemed an intriguing idea; and as my mental computer clicked around, I fed it questions like, "Thailand? or tin cans?" "For moppets? or for mothers?" And the answer (as some of you already know) came out: Tincraft for Christmas, *New York, William Morrow, 1969.*

Now, the moral of that story is one that you have heard before, namely: never underestimate the power of a woman—especially if her name is Helen Van Pelt Wilson. Because, by beguiling, belaying, and beleaguering, she has wrested not one, but two books on tincraft from a most un-canny source, namely me.

Which brings me to the burden of my song:

The Case for Tin Cans

Up until the time I was hog-tied by Helen, I'd thought very little about tin cans. Once, for a house tour featuring the arts, I'd been called upon to make wilt-proof symbols from sheet tin: an inkwell with a plumed pen, representing Literature, an artist's palette and brush for Painting, etcetera. But while that (providentially) occasioned the purchase of a pair of superior snips, it did not put me face to face with the properties, and what I have since discovered to be the almost limitless possibilities, of tin cans.

Let me first list a few of their properties. Cans are bright, beautiful, crisp. They are surprisingly durable, but conveniently degradable. They are therefore both economical and ecological. At the same time they are elegant. They are practical if only because they are plentiful, but perhaps most appealing of all, they are free and they're fun.

As for the possibilities, come with me.

Lucy Sargent

Westport, Connecticut
May 1972

contents

PART ONE
tools, materials, and techniques

WORK A LITTLE MAGIC

I can hear you exclaim, "Oh, I could *never* do that! That's *much* too complicated!"

But may I tell you a secret? It only *looks* complicated.

"That's what *you* say!" you say.

But, believe me, it's true. Let me tell you why.

Tin *curls* when you cut it. As *you* cut straight ahead, *it* spirals off, all on its own, in extravagant, rococo curves. And there *you* are, getting all the credit for producing an elaborate effect, when you have done next to nothing at all, *the tin has done it for you!*

Take the Frosty Star, page 120, for example. Have you ever beheld such an incredible collection of curls?

Now look at the diagram and see how elementary it is: Just a series of straight lines, that's all. A good *many* straight lines, perhaps, but nothing complicated, nothing you have to *think* about, nothing that requires special talent or skill. All that you *must* have is the desire to make it.

That is true of most of the projects in this book. For though I like to think there is something here for everyone, including the accomplished craftsman, I have consciously kept the techniques simple. After all, this is meant to be fun! For, Andy Warhol notwithstanding, you and I would not be working with tin cans if we had any serious pretensions as artists. But the greatest fun comes from transforming a can so completely that, as your friends stand there gasping at your achievement, you can afford to say with unconcealed glee, "Campbell's Soo-oop!" . . . "Cat foo-ood!"

They will call you a genius, but you will know that it's just a little magic . . . a simple sleight of hand . . . a *trompe l'oeil*, as it were . . . because of a few tricks noted down in this book.

TIPS ON TIN CANS

Would you believe that the annual consumption of cans in America averages 389 per *person?* Yet most people never look at them twice; never even notice those gorgeous gold linings, let alone distinguish among them.

To get the current facts straight, 91 percent of the cans manufactured in this country are made of sheet *steel*, on some of which there may be a layer of tin, a ten-thousandth of an inch thick, but on many of which there is no tin at all. Quite intentionally, our manufacturers have devised and substituted lacquers as protective coatings, not only to improve the product, but to avoid dependence on an imported commodity. This works out to our advantage, yours and mine, because it gives us ten different gauges (weights) and fifteen luscious lacquers to choose from.

For ornaments and jewelry, my favorite can is Campbell's Soup—any kind—but especially Beef Bouillon, Black Bean, Golden Mushroom, Green Pea, Hot Dog Bean, Onion, and Pepper Pot. They are all made from a lovely, limber grade of frosted tin, lacquered in exquisite shades of gold, varying from pearly Early Mediterranean to rich Florentine Renaissance. As if that weren't enough, the outsides of the cans are often marked in squares, diamonds, and even pentagons, making them ideal for belts or necklaces, with all the measuring done for you (see Sunburst Chain and Earrings, page 165).

My second favorite is the V-8 Juice can, especially the largest size, from which I have made many of my stars. It has not the reddest, but the richest, gold lining and comes in a proportionately impressive gauge of tin sheet, altogether stunning and refined.

When shopping, bear in mind that all *red* fruits and vegetables require lacquered linings. For different reasons, so do corn, peas, and fish, not to mention oddments like ripe olives, mushrooms, sweet potatoes, and tortillas.

Then, there is the whole category of commercial cans to choose from. Approach your fish market for the large, lightweight, all-silver cans in which fresh fish are delivered. My man, Ralph DeSio, disposes of his cans once a week, and if I get there by ten on Saturday morning, I go home loaded.

Mildred and John Wallahora, good neighbors and proprietors of our Country Store, save all their imported ham cans for me, and even collect huge four- and five-gallon cans (which contain everything from fresh blueberries to soy bean oil) from the Three Bears Restaurant across the way. When I was making the Canapear Curtain, every last person in the Wallahora enterprise drank Campbell's Soup for lunch, day after day, to meet "our" deadline. This kind of friendship makes all things possible, and I am sure that if you seek it out, the same good fortune will be yours.

TOOLS AND MATERIALS

The tricks of the trade, or secrets, as craftsmen have called them for centuries, are really little more than techniques facilitated by proper tools. Happily, the tools for tincraft are mostly things you already have around the house, such as a screwdriver, pliers, kitchen chopping bowl, awl, ruler, gloves.

THE SNIPS

But the one tool on the list that you can't get along without, the one tool that makes all the difference between joy and despair, pain and pleasure, is the snips. Snips you really should treat yourself to. The Klenk Aviation Snips are my favorites, both for the beautifully milled blades and the easy-grip handles. They cut straight and in circles; come in both right- and left-handed models; and their double cam action makes the tin seem as soft as butter.

Plain-edged snips. These make a perfectly smooth cut and should be used for all Early American designs and most utilitarian objects, such as cookie cutters and cachepots.

Finely-serrated snips. There are 36 tiny teeth on the blades, which produce an exquisitely beaded edge on the tin, very nice for jewelry in particular and ornaments generally.

Coarsely-serrated snips. These have fewer, but more pronounced, teeth on the blades, and produce a stunning coruscation along the edge of the tin. They are essential for any project involving a Frosty Star.

In my earlier book, *Tincraft for Christmas,* I said that if you were to have only one pair of snips, perhaps the finely-serrated would be the one to choose, but as I write now, I realize that someone "borrowed" my pair, and I have produced this entire book without them, relying on the plain-edged snips for most designs and the coarsely-serrated for special ones.

Perhaps the best way to make a choice among the snips, if you *must* make a choice, is to pick out your favorite designs and see what they call for. I always specify which snips to use *if it matters.* Otherwise, I just say *snips.* (You will find sources of supply listed at the end of this book.)

OTHER ESSENTIAL TOOLS

Awl. Used to inscribe lines and make holes. Even though in many instances you could substitute a nail for the awl, you will have such repeated use for this tool, it is worth buying.

Ball-pein hammer. A round-headed, flat-ended hammer is indispensable. The head should be perfectly round, *no point at all.* Eight ounces is a comfortable weight.

Can opener. Unfortunately, not all wall can openers remove rims. My new Can-o-mat Series DL 245 S2 does, but to make absolutely certain that yours will, I suggest you take a can with you to the hardware store and try out the openers before buying one.

Hammering board. A *hardwood* board is best for striking clean, neat awl holes and for hammering. You would substitute a *softwood* board only for piercing a Paul Revere lantern, when a deeper textural effect is desired.

Pliers. You will need *long-nosed* pliers (or needle-nosed), which are flat on the inside, and *round-nosed* pliers, which are completely round. The longer and narrower you can find them, the better.

There are other tools and materials that you will

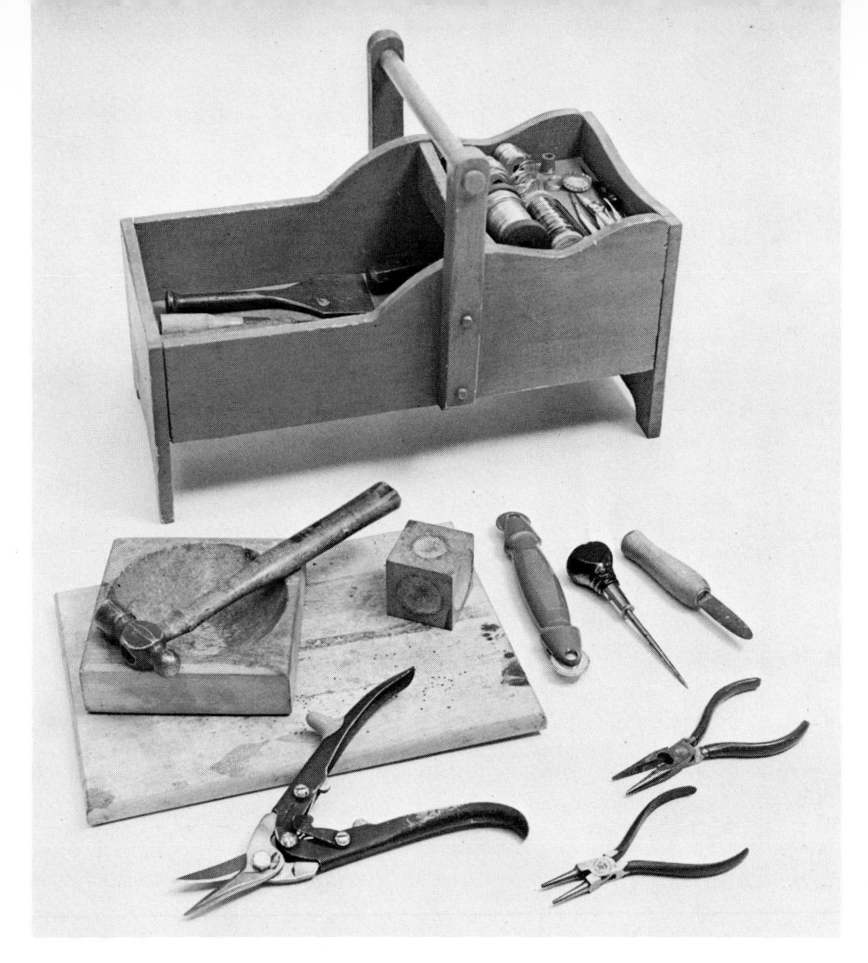

TOOLS FOR TINCRAFTING

(front row) Klenk Aviation Snips, round-nosed pliers, long-nosed pliers.

(middle row) Hardwood hammering board, kitchen chopping bowl, ball-pein hammer, dapping block, screen installation tool, awl, oyster opener.

(back row) Handy tool chest, not essential.

have use for from time to time, and I list them here, together with some of the brands which I have found particularly satisfactory, always acknowledging the fact that substitutions can, and sometimes must, be made.

ADDITIONAL SUPPLIES

Architect's Linen. This is a transparent, finely woven, glazed cloth, far more durable and desirable for making patterns than tracing paper.

Chisels. You may wish to use both the narrow, sharp wood chisel and the broader, heavier cold chisel (see photo #2) to ornament tin.

Chopping bowl or dapping block. You will use this with the ball-pein hammer to curve surfaces. A deeply rounded wooden bowl is satisfactory for most projects, except when you need to raise *only a portion* of a surface; then you must have the dapping block.

Compass. An inexpensive compass is perfectly adequate to inscribe a line around the side of can when cutting it down, or to find the center of a lid.

Glues. *Weldit Cement* is still the best all-purpose clear cement I've found. While it holds up well under pressure, the bond *can* be broken, which is *sometimes* an advantage. When doing fine work, apply Weldit with a toothpick. *Epoxy 220* is excellent but expensive and a nuisance to mix. Because it is "forever," apply it only when you are sure your design is right, or when it is subject to sudden strains, as in the Dinkybird Diner. *Rubber cement* is the *only* cement to use for applying patterns because it rubs off with the fingers, leaving a clean surface.

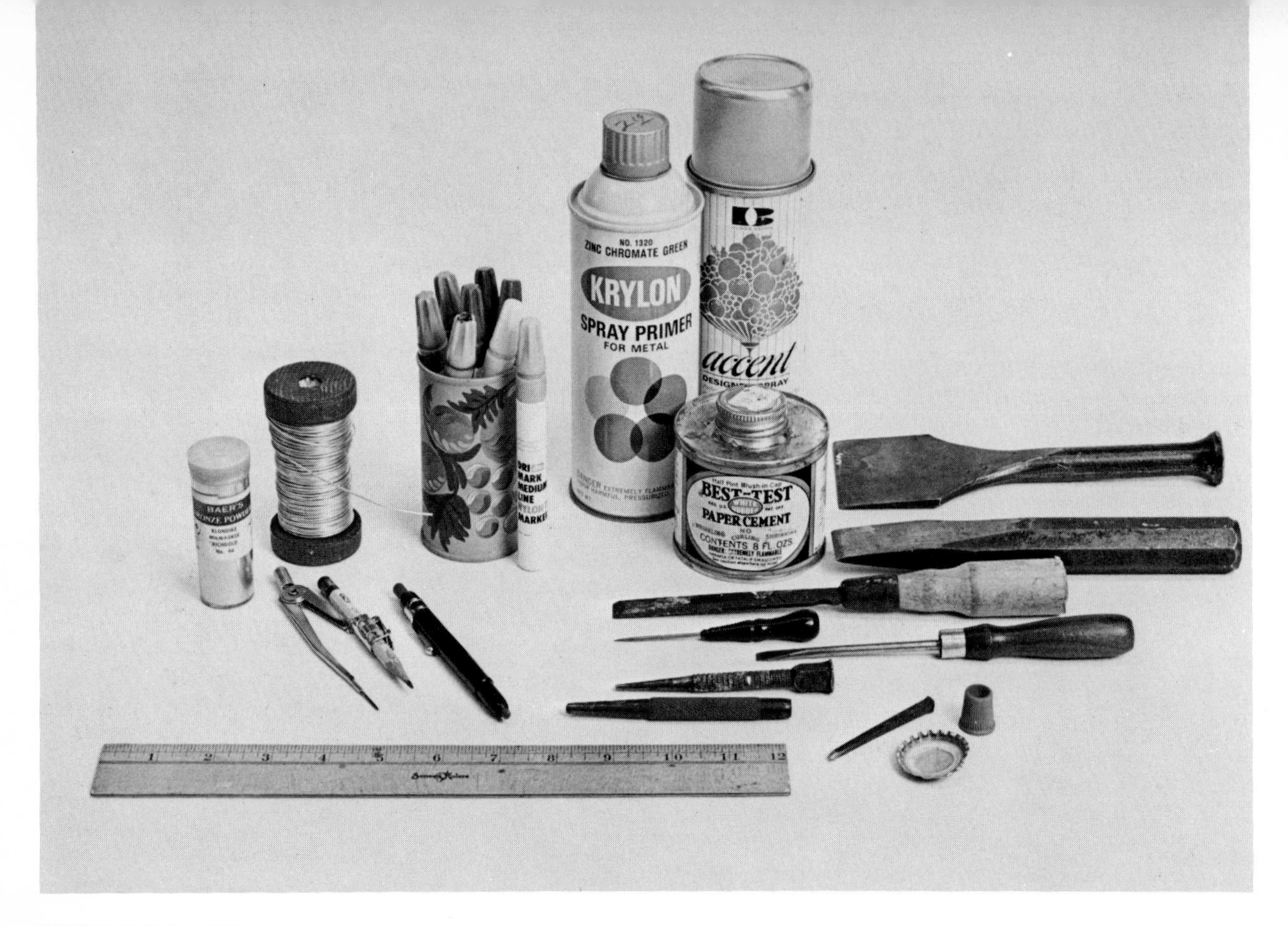

MISCELLANEOUS TOOLS

(back row, left to right) Baer's Bronze Powder, Nu-Gold (or jeweler's bronze wire), Dri-Mark pens, Krylon zinc chromate metal primer, Illinois Bronze *Accent* tôle red spray, Best-Test rubber cement, two cold chisels.

(front row, left to right) Ruler, compass, grease pencil, two nail sets, stiletto, wood chisel, screw-driver, sawed-off nail, beer bottle cap, and toothpaste cap.

Grease pencil. A marker readily rubbed off with the fingers; the red screw-type is preferable. A felt-tip pen can also be used and markings eradicated with turpentine, paint thinner, or nail polish remover.

Knives. You may have use for two kitchen knives: a single-bladed, oval-tipped *oyster knife,* the best substitute for the Early American chisels used to make the slots in pierced lanterns; and a sturdy *paring knife* to open a can amidships, as in the Mexican Sun Mirror, page 118.

Lithopone. A fine chalky substance, available in most drugstores, is often used in preference to carbon paper in making tracings on tin.

Metal hammering block. Use when flattening round wire, as in the Florentine Chain (page 167) and the Cocktail Picks (page 144).

Paints and brushes. For quick decorative effects I recommend *Dri-Mark (oil-based) permanent markers.* They come in three tip-sizes and twelve gay colors, and can be made fairly permanent if thoroughly dry before being sprayed with clear acrylic. *Talens Transparent Glass Paints* 190B10 is an excellent, inexpensive kit providing nine rich colors, which can even be baked in the oven and thereby be made virtually indestructible. They are ideal for "treasures." As for spray paints, both *Krylon* and *Illinois Bronze* are excellent brands. The best olive green is *Krylon Zinc Chromate Metal Primer;* the best Early American barn red is *Illinois Bronze Tôle Red Primer.* Use *Krylon Crystal Clear Acrylic* for a final protective coat on all treasures. Artists' oil paints can be made durable and relatively quick-drying when mixed with *Pratt and Lambert's 61 Varnish* (gloss or satin, depending on the finish you want). For bronzing, *Giles' Black Varnish* is best. As for brushes, *#0 Fashion Design* is excellent for all fine work, especially border lines. For general work, any pointed plump brush will do.

Screen installation tool. Two wheels at either end of a handle (one convex, the other concave), make this the best hand tool for inscribing simple single borders on sconces or veins on large leaves. Work the convex wheel from the front, the concave from the back.

Steel wool. Kitchen soap pads are fine for general cleaning and burnishing purposes, but *0000,* which is almost as soft as absorbent cotton, produces a lovely satin finish.

Upholsterer's tacks. I use Decora No. R 148 J exclusively. If they come overly antiqued, clean them up with steel wool.

Vise. A small inexpensive vise is adequate for making scrolls and holding pieces for soldering.

Wire. You will have use for all sorts of wire, from the heavy 10-gauge to the very fine hair wire, made from steel, copper, and brass, in straight lengths, coils, or spools, obtainable at florists', hardware stores, or welding shops. *Nu-Gold,* or jeweler's bronze wire, is good for jewelry. *Sash cord wire,* made of 32 twisted fine steel strands, is good for stems of flowers having many stamens, such as roses.

Asbestos mat. A 2′ x 3′ piece of ¼″-thick asbestos board is preferable, but a thin roll of asbestos tacked onto a ¼″ plywood board could be substituted.

Propane torch. I have used only the Bernz-O-matic, which I have found entirely satisfactory.

Dunton's Tinner's Fluid. A mild solution of hydrochloric acid, it cleans the tin when heated by the torch.

50/50 solid core solder. Kester's fine 50/50 is the most desirable but frequently hard to find. Any solid core *(no rosin or acid core)* will do.

Brush. Cheap dime-store watercolor brushes will do for applying the tinner's fluid.

TECHNIQUES

To answer *your* question first: No, I do not cut myself. Tin is so predictable once you become familiar with it—which doesn't take long—that you will quickly learn to pick it up *lightly*; and as you work, hold it *firmly*. You will be surprised at how soft and flexible it is, unless you cut short, stubby, jagged points, in which case, look out!

Little hairlike "hangnails," which you are also apt to make as you struggle to open out a can, are sharp; but you will soon learn to avoid them, or even not to make them, by keeping your snips deep in the cut, and *never* withdrawing, no matter what.

But, if you do get pricked on a point, you will probably find it smarts less than a paper cut, and a Band-Aid is the most I've ever required even with all my experimentation.

I have found only three situations in which you might really want to wear gloves: when you cut along the seam to open up the can; when you cut along a curved, ribbed section in which you are determined *not* to put a bend; and when you reshape the cone and sides of a pierced lantern. Otherwise, you'll probably find that gloves are more hindrance than help, but if you are nervous without them, by all means use them.

To answer Barbara Walters' question on the Today Show: The most difficult thing to do in tincraft is to open out the can in order to use it. *If you can do that, you can do anything!* When she put the question to me, I assumed that she was referring to designs rather than situations, and I hesitated because I honestly had trouble thinking of anything really difficult in all of *Tincraft for Christmas.* But, slugging your way along the seam of any can larger than a soup can is rough and disagreeable—a fact I steadfastly ignore because *I want that piece of tin!*

But let's start at the beginning.

POLISHING CANS

Presumably the can has been washed and dried, a process I mention only because it seems to desiccate many of the label-glues and make it easier to pare the labels off with a knife. Otherwise they seem inclined to stick. Some stick regardless, and not even paint remover helps. Those cans you may just as well abandon.

Run the cans under hot water to open the pores, and polish with soapy steel wool. Use only a soapy cloth on lacquers, lest you mar their perfection.

There are cans coming through now with lithographing that no paint remover can touch. I am told they have been baked on, but why I don't know. Except for the Marshall Kippered Herring can, which is essential for the Pineapple Wall Pockets, I don't bother with them. There are plenty of other cans that are pure perfection.

REMOVING RIMS

In order to obtain a flat piece of tin from a round can, it is best to remove the top and bottom lids and both rims. When neatly cut with a wall can opener, the lids and rims can be used in many ways.

To remove rims, put the can into the opener *sidewise,* catching the rim under the wheel just as you would if opening it from the top. Turn the handle and off it comes . . . with *some* can openers, that is! Unfortunately, not all can openers will remove rims, and even those that will cannot accommodate the large commercial cans. So, for ordinary, everyday cans, try out the opener before buying it; and for the outsized cans, ask your grocer (or whoever gave them to you—restaurant or delicatessen owner, etc.) if he will be kind enough to remove the rims for you.

The old-fashioned hand opener (which you jab into the can and saw up and down with) will, on cans whose rims are made with a single fold of tin rather than a double, produce a decorative edge *as it cleaves the fold.* (Notice the band under the lunette on the Mexican *Retablo,* page 124) But otherwise, hand can openers—even the improved varieties—are too dull, or too shallow, to remove the lids and rims of outsized cans.

But there is still another way of opening a can with a paring knife (obviously not your best one, but a good stiff one). Holding the can firmly with one hand, plunge the knife into it near the seam; then cut around the can as you would a round loaf of Boston brown bread. You'll be surprised how smooth and straight the cut will be, not a bit jagged. It's a nifty way to get the whole end off a can if your present can opener won't do it. I used this technique on the Mexican Sun Mirror, where I wanted three sets of rays of different depths.

OPENING OUT CANS AND CUTTING CURVED RIBS

You will probably want to wear gloves for this.

Since the seam is rigid, and cannot curl up from the snip as it normally does, make the cut along the *left side of the seam* so that the more flexible body of the can will give a little and ease your struggle. In order to avoid sliver-y "hangnails," keep your snips *deep in the cut* and forge ever-forward, bracing them against your middle.

If you are planning to use the sides of the can *flat,* pull the can open and step on it. Hammer it first on the *convex* side; then turn it over and hammer it on the opposite side, switching back and forth until the piece is as flat as you want it.

If, however, you need *curved ribs,* as in the Festive Swag, pull the can open cautiously to avoid putting a crease in them, and, holding the can *concave side up,* cut along the row of ribs (on the right side of the can if you are right-handed; on the left if left-handed).

The *plain* strips on the outer edges of the can will give you no trouble, but the rows of ribs will, because they are rigid. The nut on the snips will conflict with the curve and *almost* force it out of shape, but not if you're careful. Keep the nut on top of the rib. If the snips slip out of position, replace them carefully before continuing to cut, to avoid making tiny hangnails. Do *not* attempt to come back from the far side to meet the cut halfway: there is too much rigid metal involved, and you will get a jagged cut. Fortunately, the *fine-ribbed* cans you are most apt to use are made from a fairly light grade of tin that is more flexible than, say, a V-8 Juice can.

CUTTING DOWN CANS

Every now and again, when a can is too tall for your purposes, you will have to cut it down.

In order to make the new line even all the way around, open your compass to the required depth. Place the *pencil* end against the rim of the can. Using the rim as a guide, push the compass around the sides of the can, scoring the new line with the *sharp* end.

Bear down on the sharp end, but keep your eye on the pencil end, to be sure that it hugs the rim. If you were to reverse the compass position, you would discover that the sharp end would catch on the rim, and the pencil end would frequently leave no mark at all on the can side.

Working along the *left* side of the seam, cut down to the line in an *easy curve,* and continue on around.

STRIKING AWL HOLES

To make a neat awl hole, always use a *hardwood* board. Strike the hole from the front. Strike it again from the back. Hammer it flat *from the back.*

IMPORTANCE OF HAMMERING

Hammering is vital to the smooth execution of almost every design in this book. It is essential in flattening a piece of tin for cutting, especially when a pattern must be applied; it is essential for smoothing and softening the edges which have just been cut; it is essential for giving a design depth and dimension. Hammering makes the metal supple and manageable. Every project in the book contains specific instructions on how to use the hammer.

Tin *comes up to meet the flat of the hammer;* you can therefore make a piece *curve up* by hammering it *from one side only,* or make it *perfectly flat* by hammering it *from both sides.*

Even more decisive curves can be made with many tiny strokes of the *round* end of the ball-pein hammer. If used on a flat board, it will produce a shallow cup; if used in a chopping bowl, a much deeper cup, even a dish if you persist; and if used in a dapping block, an *interior area* can be *raised against a flat ground.* (See Two Easy, Easy Sconces, page 84.) Flowers and leaves almost appear to breathe when beaten with the ball-pein on a flat board.

EFFECT OF OTHER TOOLS UPON TIN

Those of you who are right-handed may have noticed that tin curls *down from the snips on the right,* and *up from the snips on the left,* as you cut a strip. The finer the strip you cut, the more it will curl. (That is the secret of the Frosty Star.) If you have hammered the tin well, and hold the snips *vertical* to the tin, you should get perfect curls.

If for some reason, however, you do get a kink in a curl, you can straighten it with the long-nosed pliers and recurl it, either with your fingers or, in a tight situation, with the round-nosed pliers. Preferring easy curves to corkscrew curls, I usually use my fingers to encourage a curve and twist my wrist to make it spiral a bit, and in that way avoid the "bends." (See the Star of Bethlehem, page 180) You will discover that your fingers make very sensitive tools—pinching, stroking, almost kneading the metal.

FRINGING TIN

Just as you might think, fringing is the cutting of straight, narrow strips along, or around, the edge of a design. The cuts may vary in depth, depending on the project, but usually not in width, being nearly always between 1/16″ and 1/8″ wide.

Sometimes you will want the fringe flat, in which case hammer it from the *base* out to the tips. If the fringe is quite deep, as with the Crazy Chrysanthemum, run the flat of the hammer up into the curls to straighten them a bit before hammering systematically from the base. You can prevent the strands from curling to some extent by holding the fingers of your helping hand in their way as they curl down.

ORNAMENTING RIMS

Like fringing, ornamenting a rim involves a series of straight cuts through the raw edge left on the rim as it comes off the can opener, each cut no more than 1/8″ from the next, working around the rim counterclockwise. This will turn the metal diagonally and create a most pleasing pattern, as you can see, for instance, in the Evening Star Mirror, page 122.

DIVIDING SIDES OF CANS INTO EQUAL SECTIONS

The simplest way to divide the sides of a can into equal sections is to cut a sheet of paper to fit *exactly around the sides, within the rims, butting perfectly.* Fold it in half, three or four times, *creasing it sharply.* (The number of folds will depend on the project.) Coat the sides of the can with rubber cement; lay one end of the creased paper *along the seam*—or *centered on the seam,* as the case may be—and smooth it around the can. Cut along the creases.

Typing paper, shelf paper, even tissue paper are all good; but *not* newspaper.

DIVIDING LIDS INTO EQUAL SECTIONS

Until your eye becomes accustomed to judging, you may want to find the center of a lid with the compass, marking the spot with grease pencil or felt-tip pen.

Holding the lid in your helping hand and facing it squarely, cut up to within the specified distance from the center dot.

Turn the lid around so the cut is at 12 o'clock, and come up from 6 o'clock to meet it.

Turn the lid so the cuts are at 9 and 3 o'clock, and come up from 6 o'clock again. Do the same on the opposite side.

Continue dividing the sections in half, *always facing the lid squarely* so that you can judge the midpoint accurately.

DIVIDING SIDES OF CANS INTO EQUAL SECTIONS

TRACING PATTERNS

If you plan to use a pattern only once, tracing paper will do; but if you anticipate future need for it, by all means use Architect's Linen.

Lay it, *dull side up,* on the design in the book; trace the pattern and *glue the folds together* with rubber cement *before cutting out* the pattern. This will ensure a perfect mirror image when you open up the pattern.

If there is an *interior* design on the pattern, make a separate tracing of it on standard tracing paper. Moisten the back of the paper with turpentine, mineral spirits, or alcohol, and rub it with lithopone, a fine chalky substance available in most drugstores. (Lithopone is preferable to carbon paper because it is perfectly visible on the tin, yet will not affect the color of any transparent paint you might put over it.) Transfer the design to the tin with gentle pressure of the awl along the lines of the drawing.

CUTTING OUT PATTERNED DESIGNS

For best results, apply a pattern only to a *smoothly hammered* surface. Whenever possible, hold the tin *without touching the pattern,* so that the pressure of your hand will not move it.

For the same reason, work *counterclockwise* around the pattern to keep it on the *off* side of the snips (just the opposite if you are left-handed).

When maneuvering a curve, *angle* your snips; when cutting a curve to the right, *"bank" your snips to the right* and get your hand and the blades *down under the tin* as much as possible. When cutting a curve to the left, *still bank your snips to the right* but keep your hand and the blades *above the tin* as much as possible (just the opposite for left-handed people).

In a tight curve, work with the *very tips* of the snips, but on a large U-curve, *swoop* around it, using the *full length of the blades,* from deep inside the jaws to the tips, as you careen around the corner.

Pointed corners require bypassing. If you will look at the pattern for the Solomon Island Choker, page 170, and follow the arrows, you will see how you cut *into* the pointed corner, then *withdraw and bypass* the reverse curve which leads out of it, and *continue on into the next corner.* Come back later and clean up those corners, having hammered them first. You will notice, at X, that you have no alternative but to cut the curve with the snips against the pattern, but *only for the short distance from X to Y.* At Y you continue on your counterclockwise course.

Frequently in tight corners the pattern gets moved and you can't quite tell what you've cut and you know it isn't perfect. Don't worry about it. Continue on; get the whole thing cut out. Remove the pattern carefully and hammer the edges of the design flat. Then proceed to improve all the imperfect places.

VEINING

Stylized veining of leaves is done with hammer and chisels, screwdrivers, or filed-down nails.

Naturalistic veining is done either with an awl or screen installation tool. Because the screen tool is a wheel, it is easier to maneuver than the awl, but is good only for large leaves like the plume-poppy. If you need a pattern, trace it on the tin with carbon paper or lithopone. (See the following section, Pointers on Paints and Painting.) Stand up over your work, grasping the end of the tool with the *convex* wheel in both hands; then bear down hard along the lines, slowly, firmly, and accurately. Do not attempt to go over them again from the front; you will almost certainly get off the track. If the veins require extra emphasis, use the *concave* wheel from the back instead. It would be well to experiment on a piece of scrap tin first.

The awl is virtually a necessity for veining small leaves (the screen tool being too large in scale), but it may be used for large leaves also. You will, however, inscribe the lines twice from the front and not at all from the back. First hold the awl like a pencil and score along the veins slowly and firmly, just as if you were tracing the pattern. Then the second time, work back along the veins from the edge of the leaf so you can see exactly where you are going, and holding the awl like a weed digger, bear down hard, guiding the awl along the veins with the forefinger of your helping hand. Pay special attention to the points where the veins meet; they should flow together.

POINTERS ON PAINTS AND PAINTING

Some paints, such as the Talens Transparent Glass paints, are the right consistency as they come from the can, but artists' oils in tubes must be mixed with a generous amount of varnish in order to flow properly and dry within a reasonable length of time. If you want the paint shiny, use a high gloss (spar) varnish; if a matte finish is desired, use satin varnish. If you like the colors so transparent the tin shines through (as in the French Urn of Flowers, page 34), use mostly varnish (spooned into a bottle cap) and just a smidgen of pigment. If you want a more intense color, spoon the varnish onto a paper palette alongside a daub of oil paint, and mix them with the palette knife until you have the easy-flowing consistency you need. (A plain white plate makes a good palette.)

Some oil paints are naturally more transparent than others, and because I like the tin to show through, I use the following in preference: Alizarin crimson, burnt and raw umber, Indian yellow, and Prussian blue (which combined with the yellow makes a transparent green).

Titanium white is opaque, of course, and makes a very nice accent, especially in a border design. To get a *neat white dot,* use the stub end of a paintbrush or the head of a small finishing nail.

BRONZING TIN

I particularly want you to have the fun of bronzing—the gold design is so low-keyed and lovely on the silvery tin and is so easy to apply.

You will need Giles' *Black Varnish* and Baer's *Klondike Milwaukee Richgold No. 46.*

Trace off the entire design (both outline and interior lines) as accurately as possible with a finely sharpened pencil.

Moisten the back of the tracing with turpentine (or mineral spirits) and rub with a pinch of lithopone powder.

Center the design on the piece of tin and *trace only the bold outline* onto it.

With black varnish and brush, fill in the area as smoothly as possible.

Allow the varnish to become tacky (takes about ten minutes) and rub gently with the bronzing powder, applied with silk velvet or suede. Don't worry about the powder coating the tin around the design; you will wash that off later when the finished project has dried for twenty-four hours.

Position the design on the tin again and trace on the *important interior details,* such as the eye and feathers on the partridge in the Early American Match Holder (page 42), or the berries and leaves in the Easel à la Helen Hobbs (page 45). *Leave all the little repetitive etching lines to be drawn in freehand.*

Remove the tracing and very lightly etch the lithopone lines with a single-point etcher (or needle stuck in a cork), so that the black varnish shows through the gold. If you press too hard, the etcher will remove the black varnish also and you'll be right down to the tin, depriving yourself of the nice contrast between black and gold.

Let dry overnight and rinse off with water. Spray with Crystal Clear Acrylic or the gold will slowly rub off with use.

The best (and most inexpensive) book on this subject that I have found so far is Maria D. Murray's *The Art of Tray Painting.*

SCROLLS AND CANDLE CUPS

In this book, I define a scroll as a curved strip of tin used to support the candle cups of a sconce. It also becomes a hanger for wind chimes. Measuring approximetly 1⅜" x 10", the strip is reinforced by

SCROLLS

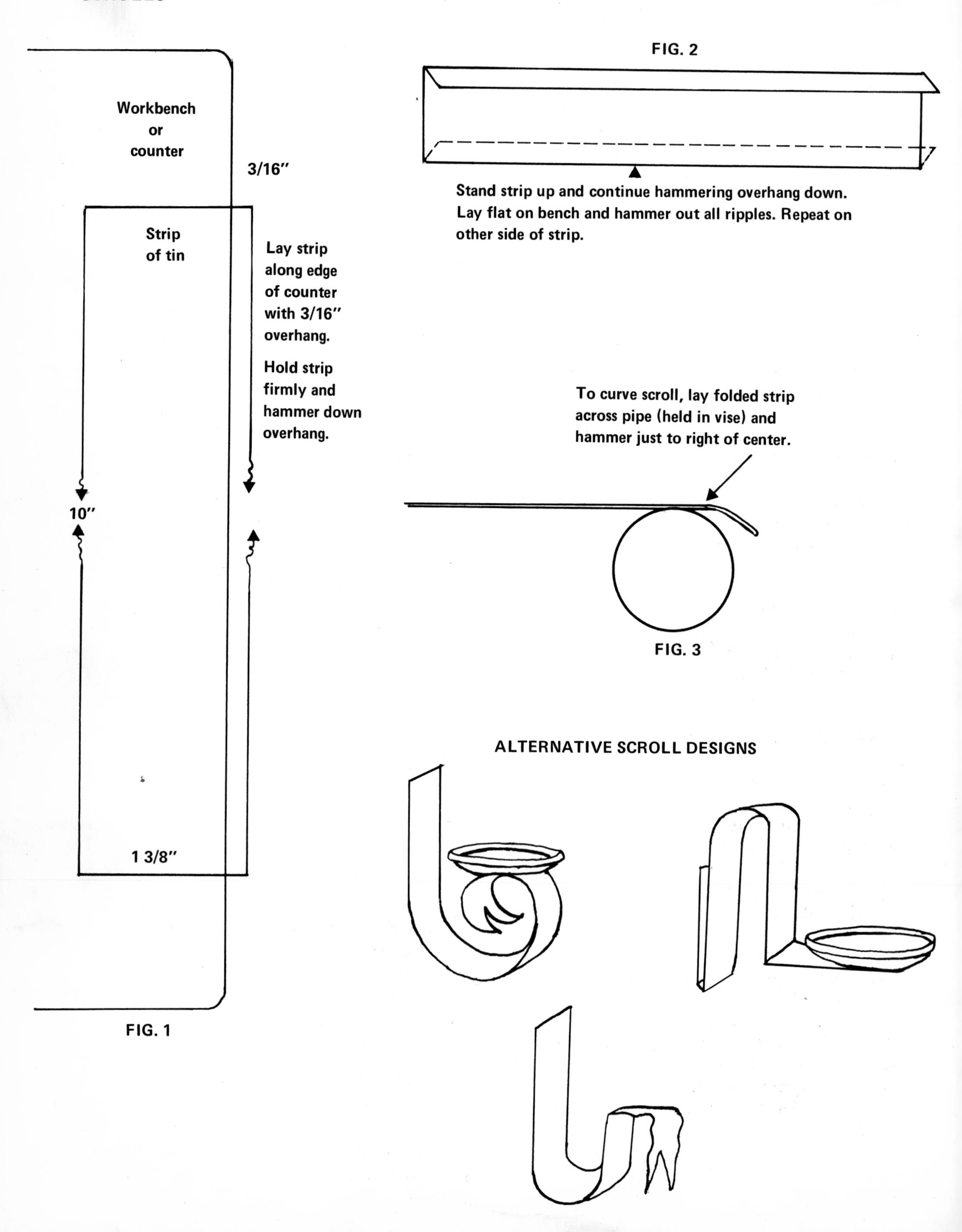
Workbench
or
counter
3/16"
Strip
of tin
Lay strip
along edge
of counter
with 3/16"
overhang.
Hold strip
firmly and
hammer down
overhang.
10"
1 3/8"
FIG. 1
FIG. 2
Stand strip up and continue hammering overhang down.
Lay flat on bench and hammer out all ripples. Repeat on
other side of strip.
To curve scroll, lay folded strip
across pipe (held in vise) and
hammer just to right of center.
FIG. 3
ALTERNATIVE SCROLL DESIGNS

narrow folds along the sides (Fig. 1). To make these folds, place the strip along a straight edge (like a counter top) so that it extends 3⁄16″ beyond. Hold it *firmly* and hammer the edge down at right angles.

Stand the strip on edge (Fig. 2) and hammer the bent portion down even farther. Then, lay it flat, and hammer out the ripples. The fold should be sharply creased. Repeat along the other side.

Cut a decorative V-notch in one end of the strip and hammer it around a broom handle or length of pipe held in a vise (Fig. 3), until you get a pleasing profile.

A candle cup can be made from the lid or whole end of a can, such as a soup can or small frozen juice can. You will use only the lid for a *crimped cup,* working around the edge of it with twists of the pliers to right and left (Fig. 1). When you've gone around twice, and the crimp is well-defined, flatten the bottom of the cup by tapping it from the back with the flat of the hammer.

For a *plain candle cup* (Fig. 2), use the whole end of the soup or juice can, so that the rim provides visual strength and finish to it. (See Removing Rims, page 14) Beat it with the round end of the ball-pein hammer in your kitchen chopping bowl until it is nicely curved.

Form the candle socket from a piece of tin 3⅛″ x 1″, curved around a broomstick or length of pipe until it overlaps about ⅛″. You will probably have to squeeze the socket well past the point of overlap *with your fingers,* and then bring it back into position by straightening any "bends" in it with the *long-nosed pliers.* You can form a perfect cylinder by using the long-nosed pliers in this fashion. (Note the variation used in the Pineapple Sconce, page 96.)

The candle socket can be joined together, and to the candle cup, with either glue or solder.

CANDLE CUPS

PLAIN CUP

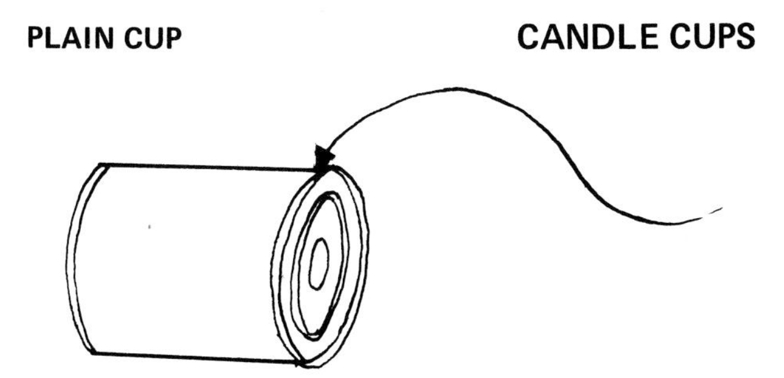

Remove whole end of can including rim, by putting rim under wheel of can opener.

Beat with round end of ball-pein hammer in kitchen chopping bowl until nicely rounded.

Make candle socket from strip of tin 1" x 3 1/8", hammered around short length of pipe to overlap 1/8". Glue or solder together and to candle cup.

Some of the caps from cans of spray paint have separate center section, which allows them to accommodate both candle and flowers.

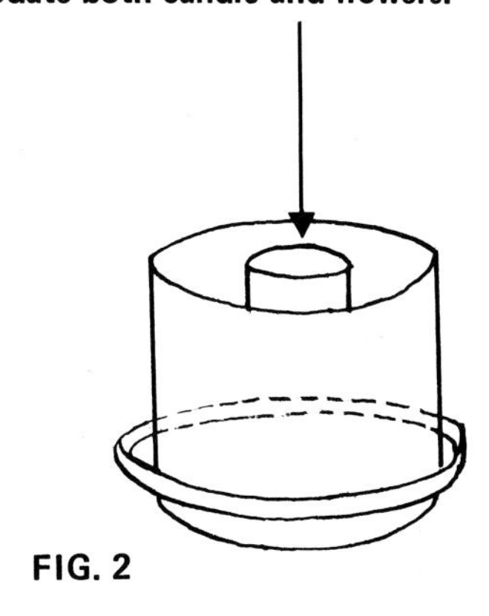

FIG. 2

CRIMPED CUP

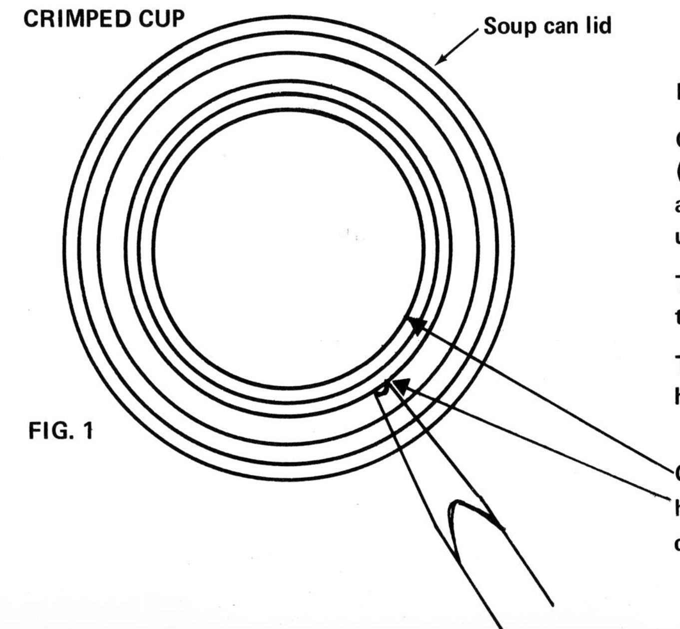

FIG. 1

Hammer lid smooth.

Grasp lid with either round or long-nosed pliers (depending upon the kind of crimp you want) and twist to right, then to left, move left; repeat until crimped all the way around.

Tighten the crimp by going around a second time close to edge.

Turn cup upside down and tap with flat of hammer to flatten base.

SOLDERING

For your own satisfaction, you really should learn how to solder. Nothing else will give you such a sense of security or craftsmanship. This attitude is not just a hangover from the Era Before Epoxy. To be sure, epoxy is as tough as solder, tougher in fact, and certainly no messier in process. But it has one big disadvantage: it can't be *un*done. Epoxy is forever, whereas a soldered joint can be reheated and reset if necessary. Besides which, solder is plenty tough enough: you can hammer your hardest without breaking the bond. Furthermore, solder has actual aesthetic appeal—a pewter-like, smooth, "caressable" character, in keeping with your tin treasure. After all, solder is 50 percent tin!

I had the worst time learning how to solder because I tried to use a gun. A gun requires the patience of Job and the dexterity of Houdini. It jiggles the joints as you try to heat them and requires building elaborate jigs to hold them steady. In addition, it takes unaccountable minutes, and who has them? My advice to you is to *forget guns, forget irons—buy yourself a propane torch.* You will have *instant success!*

Set yourself up properly. Place a 2′ x 3′ asbestos mat on your workbench (card table or whatever), preferably ¼″ thick and rigid, so that you can move it forward under the vise when necessary. (You could, instead, tack a thin roll of asbestos to a piece of ¼″plywood.)

Purchase a Bernz-O-matic (or similar) torch, a bottle of Dunton's Tinner's Fluid (a mild solution of hydrochloric acid), a spool of 50/50 (tin/lead) solder with *no* acid or rosin core in it, and a cheap watercolor brush. That's all the new equipment you will need. You already have a screwdriver, sandpaper, candle, and matches.

The secret of successful soldering is a *clean* joint, and this you can get in either of two ways: by sandpapering the parts to be soldered, or by floating them with acid *when hot.* Since it's a nuisance to sand anything unless you have to (especially *un*even surfaces), it is easier to use acid.

There is nothing to worry about. The torch will not blow up, or asphyxiate you; and the acid is too weak to do more than smart a little should it spatter (which it rarely does), in which case wash it off when you're all through. You will have to wash the soldered piece anyway (to prevent the acid from continuing to work on the metal), and you can wash your hands at the same time.

Just to show you how easy the process is, let's make some Cocktail Picks together, as on page 144. You will need an *un*lacquered can and *un*painted florist wire (#24 is about right).

Cut small oval leaves, freehand, from the side of the can. Hammer them smooth. Vein them with the awl or chisel, if you wish, and beat them from both sides with the round end of the ball-pein hammer on a flat board to make them curvaceous and supple. Hammer them flat at the stem end and place them, face down, on the asbestos mat.

Uncoil a foot or so of the florist wire and cut it into 3″ lengths. On a metal block, hammer the wires *flat* on *each* end. Place them in position halfway up the vein on the leaves.

Light your candle.

Open the bottle of acid.

Uncoil the solder so it loops over and touches the mat (see the photograph opposite).

Pick up the brush, dip it in the acid, and hold it between your first and second fingers as you would a pencil. *Keep it there throughout the entire process,* even though you will also turn on, and pick up, the propane torch with the same hand from time to time.

Pick up the screwdriver in your other hand, and *pin down* the wire stem *just below the base of a leaf.* Make sure the stem *lies flat* along the vein of the leaf so that it will make a good connection. *Keep the screwdriver there throughout the entire process and even until the solder cools.*

Turn the knob on the torch *to the left a quarter turn.* Listen! The torch is singing C Sharp above Middle C.

Hold the torch to the candle flame. Listen! Did you hear it jump two tones to E? When you hear that change in tone, you will know that the burner head is lighted, even though the flame is so low you can't see it unless you look right into it.

Listen again! Is the torch growling? If so, turn the knob a smidgen more *to the left* until it stops.

For the kind of work you are about to do, that is all the heat you will need. No big, scary flame; just steady, concentrated heat.

Hold the burner head within ¼″ of the joint you wish to solder and count . . . one second . . . two seconds . . . three seconds . . . four seconds . . . stop! Set the torch down and immediately flood the joint with acid.

Did it sizzle brightly? If not, maybe you didn't get the joint hot enough. Heat it again until the acid really sizzles.

Set the torch down. Move the coil of solder over so that its tip touches the head of the joint.

Pick up the torch and apply heat *until you see the solder melt* along the stem.

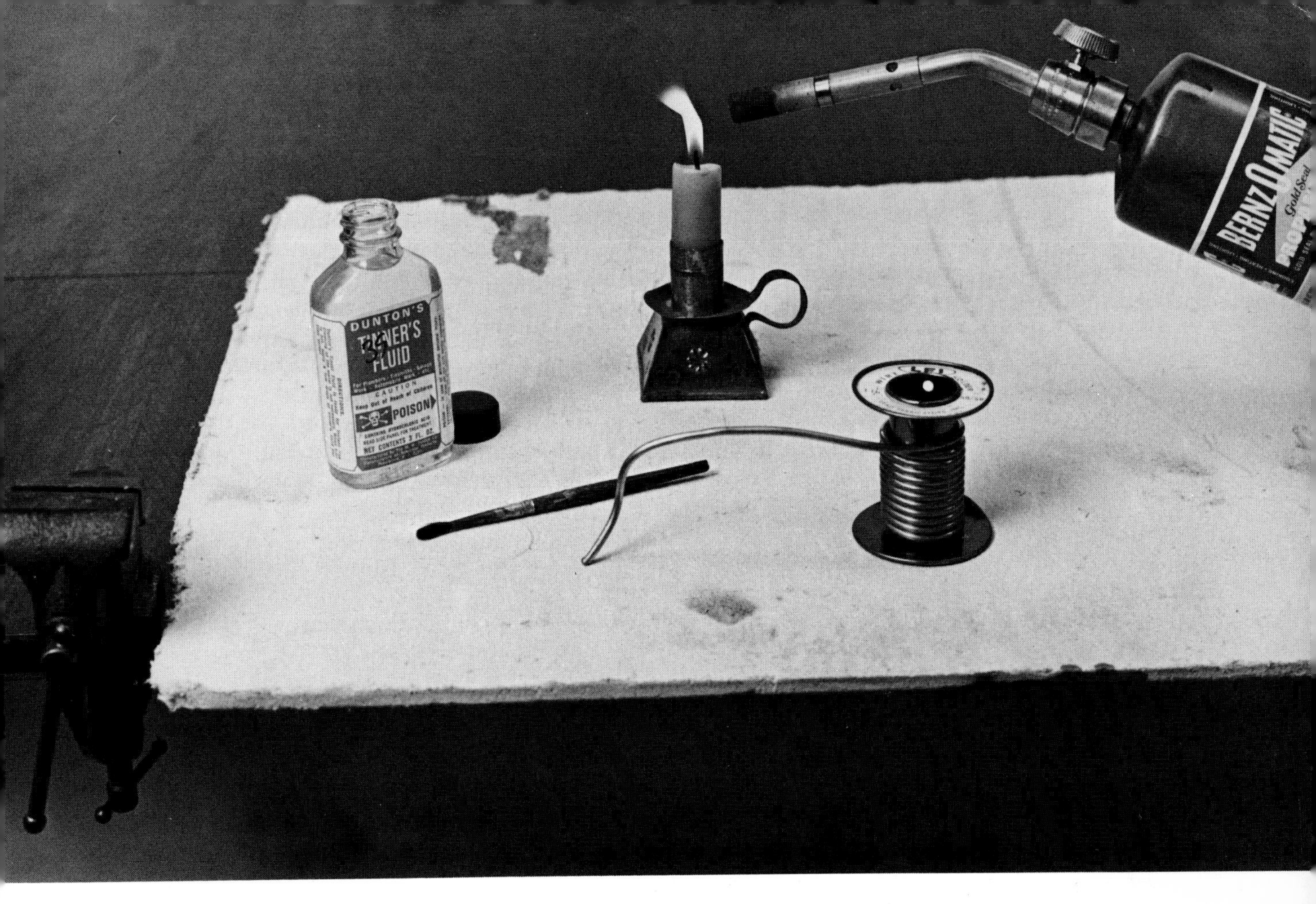

TOOLS AND MATERIALS FOR SOLDERING

(left to right) Vise, asbestos mat, Dunton's Tinner's Fluid (acid), inexpensive watercolor brush, candle, 50/50 solder, and propane torch.

Immediately knock the solder aside with the tip of the torch, without moving the screwdriver.

Keeping your eye on the joint, turn the torch off, and watch the solder cool. Remove the screwdriver.

Rinse with water and a little Arm & Hammer baking soda, if you have it.

Now, for a few alternatives. Rather than knocking the solder aside, you may prefer to cut off a ⅛" nubbin of solder and place it on the joint. If so, cut the nubbin off the coil *before* you light the torch. No point in wasting gas; a cylinder will last an amazingly long time if you conserve it. I've done this whole book on one cylinder.

On occasion you may want to hammer the nubbin of solder into a *tiny thin wafer* and slip it between the pieces to be soldered.

Other times, when you have odd-shaped pieces to be soldered together, such as a scroll and candle cup, you will have to "tin" them first. *Tinning is the coating of metal with solder.* If both pieces to be soldered are *separately* tinned where they will touch, you will then only need to apply heat in order to "marry" them.

To tin metal: apply heat; count to four; clean with acid; and melt solder on the spot. In instances when you want a thin, smooth coating, wipe the spot with steel wool while the solder is still fluid.

When soldering larger objects, you will need a larger flame, but you will rarely use the torch full-force. Too much heat burns, or oxidizes, the metal. Heat can give the metal an agreeable antique look, but, apply it a second too long and it will burn the metal black.

Once you have gained a little experience, you may decide to burn the tin *on purpose,* especially lacquered tin. The lacquers burn very easily, so you have to have perfect timing, but you can get interesting gradations of color from the original gold, to amber, to seal-brown, to black. The leaves in the French Urn of Flowers (page 34) have been shaded by partial burning. This is a tricky technique, but fun.

If you find you have to solder a lacquered surface, apply heat to the spot until the lacquer bubbles and browns; then brush it with acid to clear a space clean enough to accept solder.

As a rule, burned (or oxidized) metal will not accept solder. It has a way of looking black and greasy, and the solder "balls" off it as if it were mercury. But solder will also ball if the metal is too hot. With experimentation you will learn to differentiate between the two conditions. At times like these, it is good to have a thick wad of steel wool handy for rubbing the metal; that frequently seems to be all that is required. The steel wool is also helpful when tinning the stems of roses in a sconce. You might think it would rub the solder right off, but oddly enough, it doesn't; it just distributes it evenly and keeps it from hardening in beads along the stems.

Rinse off the acid after soldering (with bicarbonate of soda, if you have it), so it will not continue to eat the metal, and the joint will stay sleek and silvery.

PART TWO
beginner's luck

At a country fair one day, Cecily Zerega and Sally Wallace found me hammering the lids of cans into luscious apples. Intrigued, they went home to try their luck at making something round and realistic from something flat and formless, and they came up with the lively creations you will find on the next few pages.

You might say, wistfully, "They *are* clever," *implying that you are not. But then, how do you explain young Libby Zerega, who has woven a whole spider world from—what? . . . nothing but wire, a rim, and a lid. Would you say that was "beginner's luck"?*

In answer, Cecily would say that every one of us is naturally creative, in ways we may not realize, or have forgotten since our childhood. She is so convinced of this that she uses all her powers of persuasion to keep art scheduled through the formative elementary school years, so that our children will not lose their natural creative ability, and wind up like us! She also insists that if you think *something is successful, it is successful! Now if that isn't all the encouragement you need, I suggest you try your hand at some of her designs and see if you don't find yourself creating your own beginner's luck.*

Libby Zerega and her spider: Small wonder she's pleased!

THE ZEREGA ZOO

Isn't it a zany zoo? The smile on that toothy crocodile! And the backtalk the duck is giving him. And how about that busy, bushy-tailed squirrel? Every member of Cecily's zoo is as real as life, and doing his own thing.

Each animal is made from a single flat piece of tin. Glance at her patterns and see how simply she achieves the third dimension by bending legs down, jaws under, heads up; by fringing manes and tails, and rounding backs and heads with the ball-pein hammer. All quick, easy, and so successful!

TOOLS

Serrated snips
Long-nosed pliers
Ball-pein hammer and board

MATERIALS

Shiny gold and silver cans, ribbed and unribbed

The designs are all so clear and easy, you won't need any instructions other than those on the patterns themselves. Just make sure your cans are squeaky-clean.

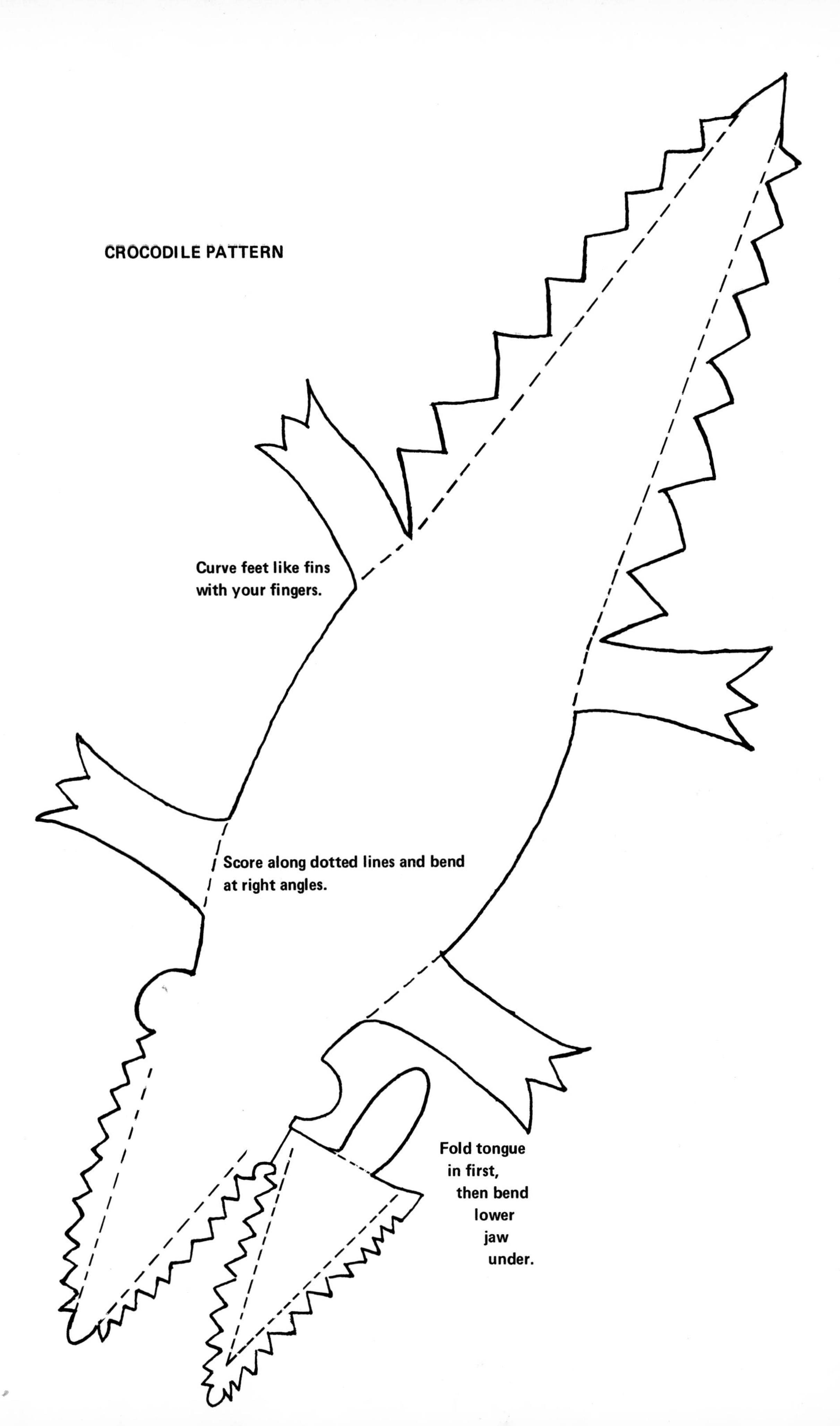
CROCODILE PATTERN
Curve feet like fins
with your fingers.
Score along dotted lines and bend
at right angles.
Fold tongue
in first,
then bend
lower
jaw
under.

DUCK PATTERN

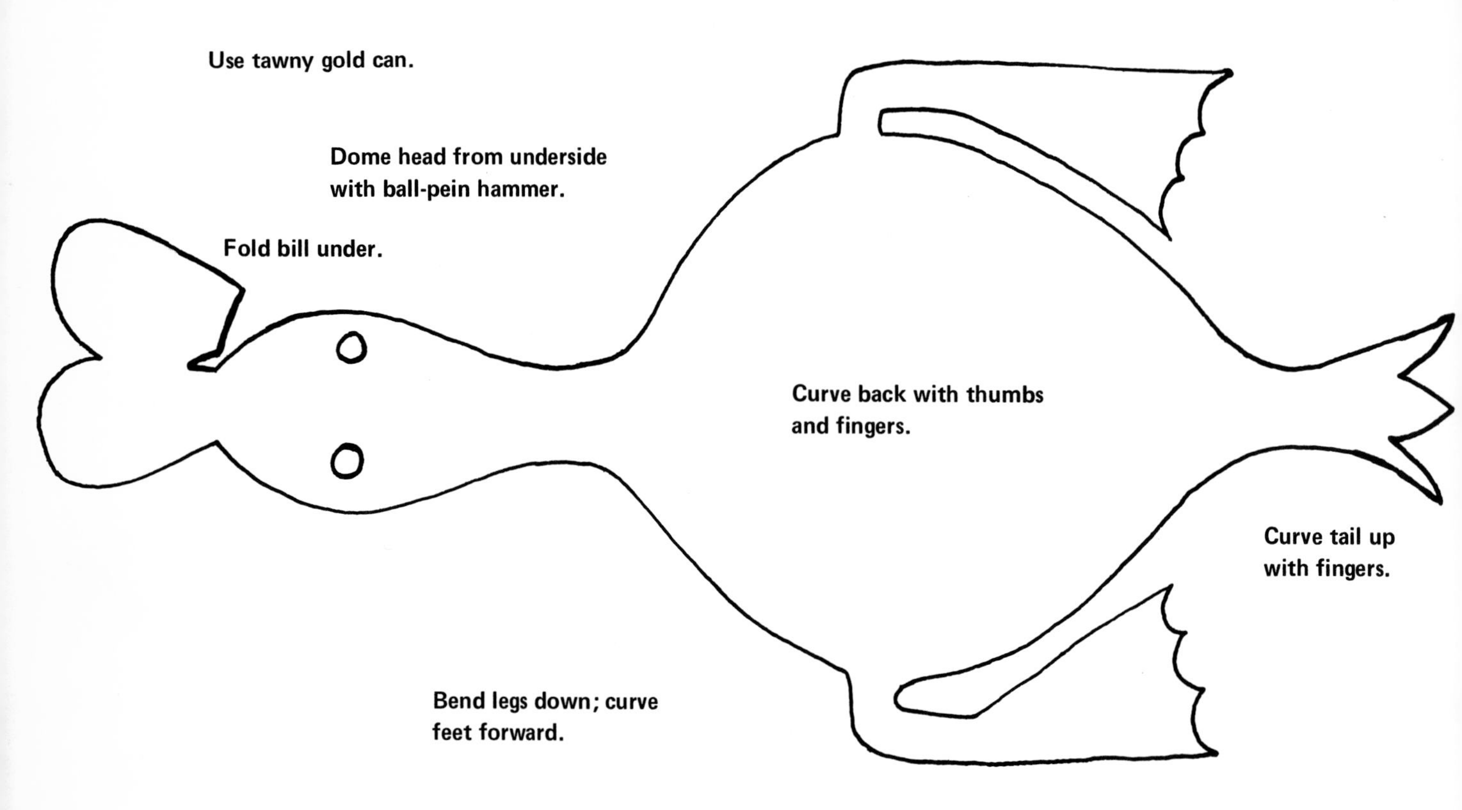

LION PATTERN

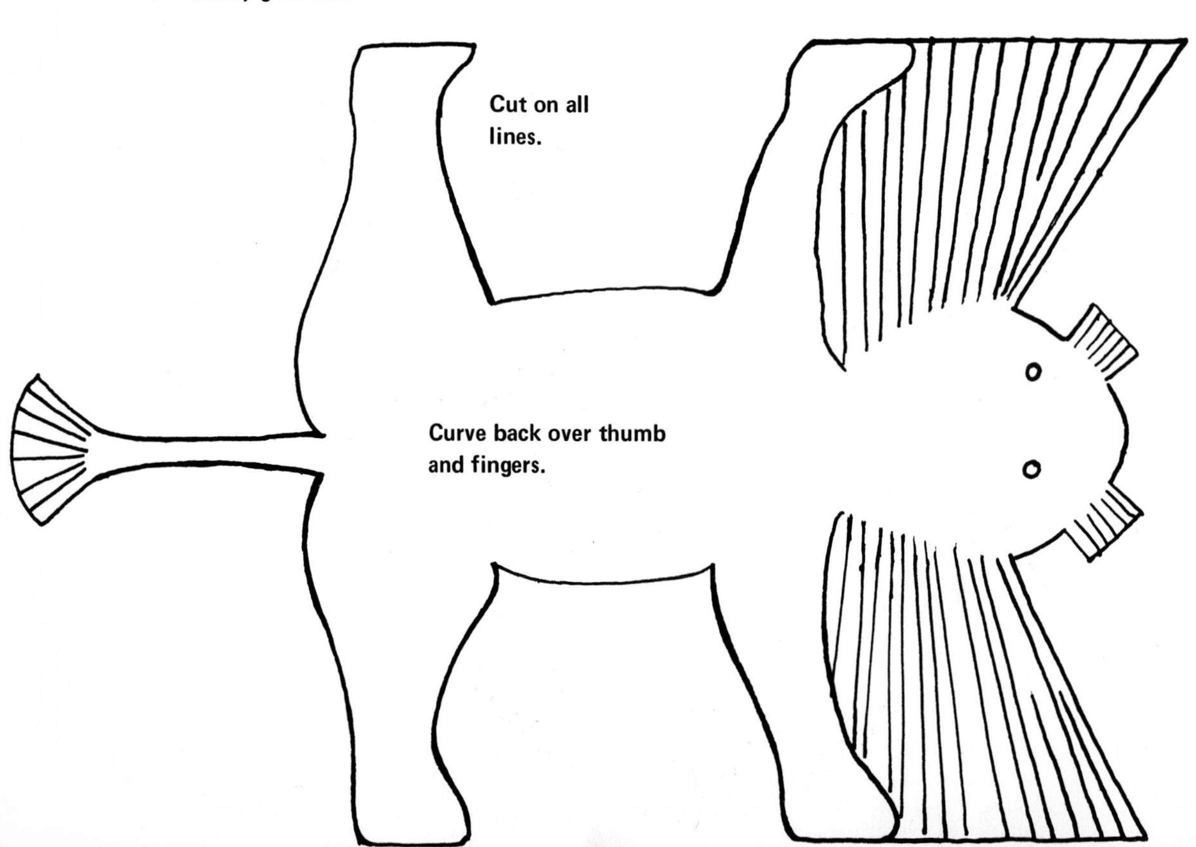

SQUIRREL PATTERN

Use a ribbed can
for this one.

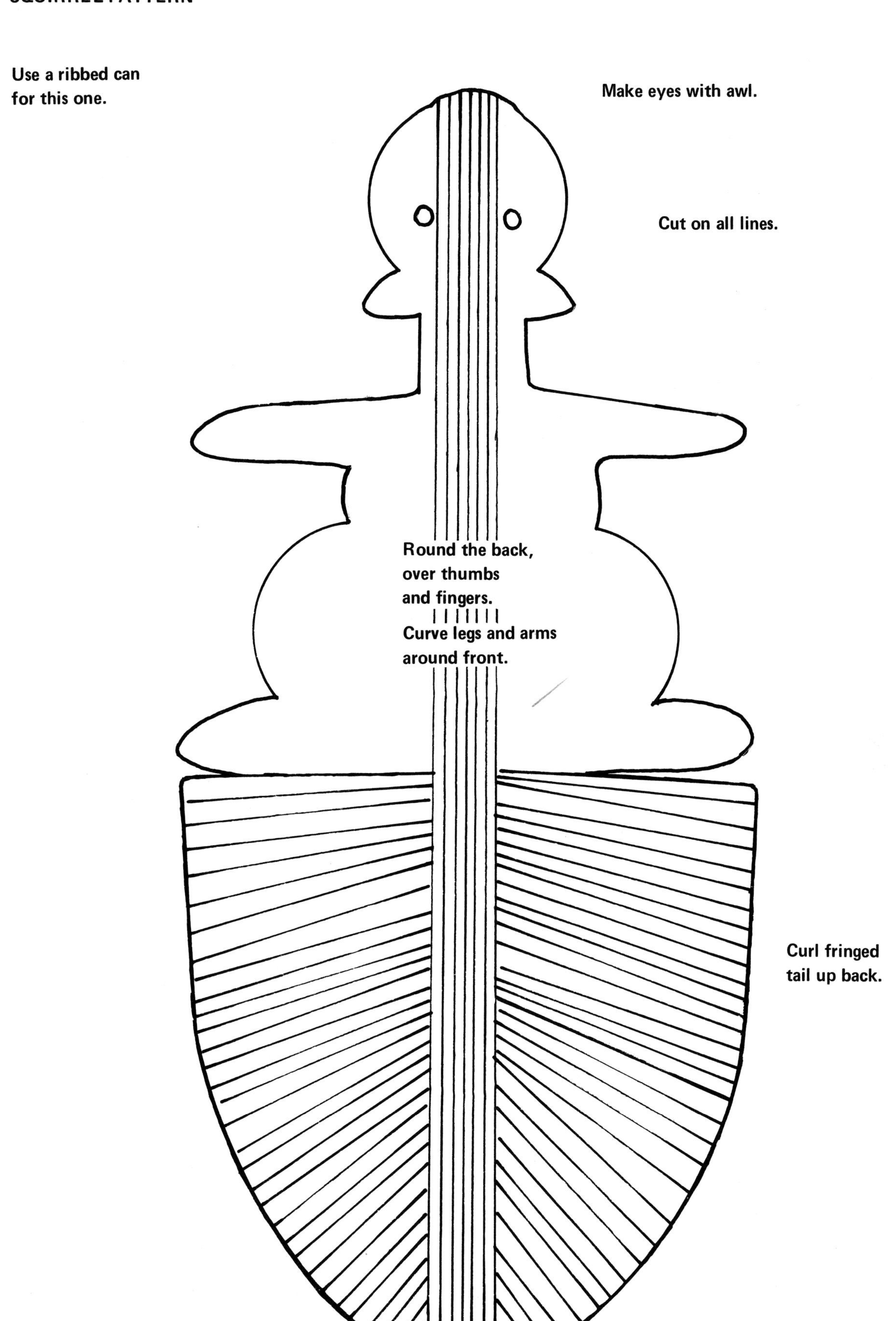

SALLY WALLACE'S PEACEABLE KINGDOM

Sally Wallace, like Cecily Zerega, is nimble-fingered and full of fancy. With no more than a glance at *Tincraft for Christmas*, she was off on her own, suiting the punishment to the crime, so to speak, by wishing a lobsterman a happy birthday with the more-than-reasonable facsimile you see on the lower left of the picture. Those strawberries, though stylized, are large as life and look twice as luscious, mounted, you notice, on the rough, "earthy" side of the clapboard. That pert bird, though made from a single flat piece of tin, is standing saucily on his own two feet, and if you lived with him, as I do, you would find him an extraordinarily "live presence." The naïveté of the cross-eyed, splay-footed lion is certainly comparable to that of Rousseau's, and the carrots . . . ! They are not only good enough to eat, they are *too* good to eat.

Sally works with the roughest of patterns; it's mostly in her head and fingers. You'll be surprised to find how much more you have in your own than you thought.

NOAH'S ARK

On the spur of the moment I turned into The Scribner Book Store one day and stumbled on an enchanting children's paperback cut-out book, *Noah and His Ark*. The story line was charming, full of lively detail; and the figures were packed with personality: Noah, the epitome of an ancient Hebrew patriarch, and Hannah, a typical sloe-eyed Semitic beauty. All the animals were there in twos, as you can see. Backing them with tin and giving them five coats of varnish makes them virtually indestructible, and is just the sort of labor of love that grandmothers were made for. Doting grandmothers, that is, and some say "dotty," too. (See color photograph, page 33)

Yes, it's time-consuming, but so are crocheted afghans, needlework chairs, and stenciled trays; but there's always the thought of those winsome blue eyes and curious pudgy fingers to cheer you on. If you'd rather, there's a version of the Christmas story, too.

TOOLS

Plain-edged snips
Long-nosed pliers
Hammer and board
Awl
Ruler
Scissors

MATERIALS

Sides of unribbed cans
Rubber cement
Spar varnish
Brush
Paper clips
Epoxy 220

All the instructions for cutting out and folding the figures are in the book itself, so that the only extra thing you have to do is to paste them on the tin with rubber cement. You will want to use the sides of *un*ribbed cans, the larger the better, of course. Preferably *un*lithographed as well, since varnishing makes the cut-out sufficiently transparent to read the label through. If you have a choice, use cans of fairly heavy weight, because you want these toys to last, after all.

You can cut out the figures before you paste them on the tin, or if you have a good supply of gallon cans, you can paste the *whole page* on, and in that way cut the figures out once instead of twice.

Round all corners.

Scoring along the lines is *very important.* Do just as instructed, *holding your awl* almost *horizontal* so the tip won't tear the paper, running it along the straight edge of the ruler. Use the *long-nosed pliers* to make all the "bends."

Glue the tabs with Epoxy 220, and secure them with paper clips until dry.

Use a brush for varnishing (dipping doesn't work) and apply two thin coats, inside and out, before sanding the edges. Stand the figures on something like a cake rack (with paper under it), or hardware cloth, and leave overnight to dry.

Apply three more coats of varnish, *lightly* sanding the rough spots between each application. That should do it, if you're diligent! And if you're feeling sorry for yourself, just remember that my tray-painting teacher would insist on ten!

I don't know whether it was worth it or not, because the ark is too small to hold all the animals anyway, but I put a bottom in it, and put hinges on the doors, so they could be opened and shut; then I screwed a handle onto the ridgepole. Perhaps the most practical way to contain it all would be to mount the ark on an old-fashioned tin tray. (See Sources of Supply, page 195)

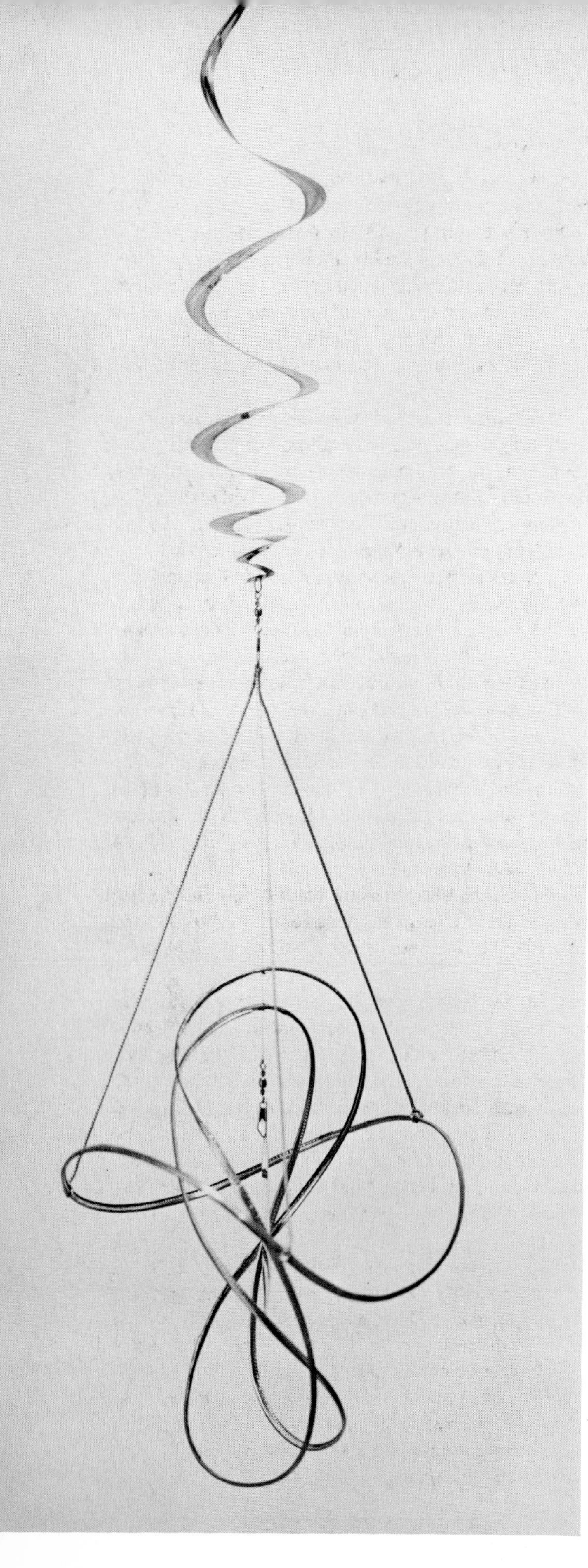

HAMOBILE

Just three rims, two lids, and gold cord, requiring perhaps an hour to assemble, but suggesting something timeless.

TOOLS

Snips
Hammer
Awl
Round-nosed pliers

MATERIALS

2 rims from a 12-pound ham can
1 rim from a 10-pound ham can
2 lids 4″ in diameter, one gold-lined for spiral hanger
Gold cord (Christmas Tie-Tie)
Gold-colored metallic sewing thread, needle, and scissors
Weldit Cement, toothpick
24-gauge florist wire
Steel wool
2 fishing swivels *(optional)*
Grease pencil or felt-tip pen *(optional)*
Krylon Bright Gold or Silver Spray Paint *(optional)*

A WORD ABOUT THE MATERIALS

Many of our American hams are being packaged in rimless plastic containers, so look for *un*lithographed cans with rims, such as the Krakus and Atalanta Polish hams. If you can find only lithographed cans, you can spray-paint the rims gold or silver.

Look also for a perfectly plain disc or lid for the center of the mobile. It somehow seems more abstract than one with concentric circles, if only because it is not so recognizably the lid of a can.

You will also find a plain disc easier to work with when cutting the spiral from which the mobile hangs, because you will have no conflicting lines to cut across.

Someone made the observation that this mobile would be even more suggestive of outer-space celestial bodies if suspended from "invisible" fish line; but having tried both, I find that the gold cord provides a pleasing visual unity between the spiral hanger and the mobile. You might, however, want to substitute fish line for the gold thread down the center. There's only one way to find out!

THE SPIRAL

Select a lid approximately 4″ in diameter with a rich gold lining. Burnish it bright with steel wool on the silver side and hammer it smooth on the *gold* side.

Working gold side up, make a small mark with grease pencil or felt-tip pen on the edge of the lid for your starting point. Make another small mark opposite the first. Make a third halfway in between.

Cutting from the first mark, gradually increase the width of the spiral until, at the *second* mark, the strip is a full ¼″ wide.

Continue increasing the spiral until, at the *third* mark, the strip measures a generous ⅜″. Continue cutting a ⅜″ strip, round and round to the center of the lid.

With the round-nosed pliers, form a hook at the beginning of the coil.

Strike an awl hole in the whorl at the opposite end; connect a fishing swivel to it with a link of florist wire.

THE SPIRAL

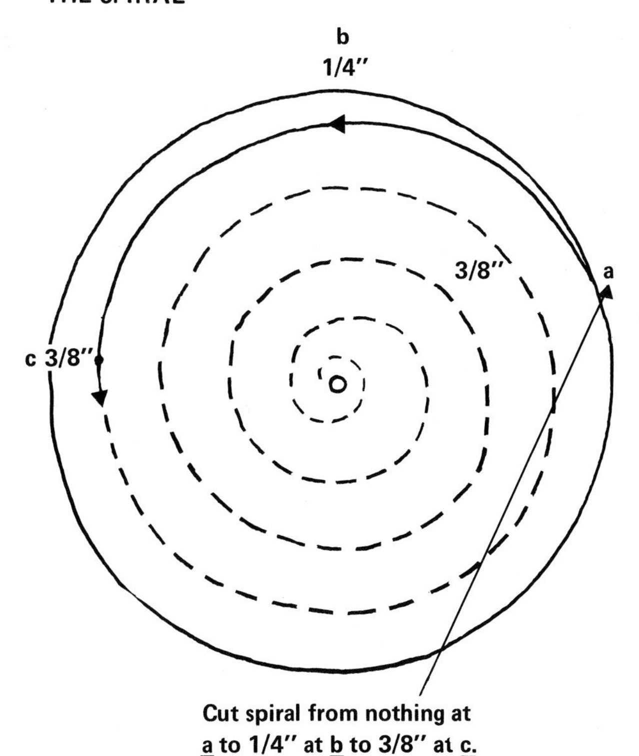

Cut spiral from nothing at a to 1/4″ at b to 3/8″ at c. Continue cutting 3/8″ strip to center.

FIG. 1

THE MOBILE

Remove the rims from the three ham cans (see Removing Rims, page 14). Snip off any conspicuously rough places and polish with steel wool. (Spray with Krylon Bright Gold or Silver if lithographed.)

Usually the rims come off the can opener in a sort of undulating, elliptical curve. *Encourage that curve* by grasping the extremities of the loop in both hands and twisting them in *opposite* directions until they resemble a loose figure 8.

The two rims (one from a 12-pound can and one from a 10-pound can) which hang *vertically, one inside the other,* should have *identical curves.* Twist them and lay them together on the table to be sure they are symmetrical. They should clear each other by ¼″ all the way around.

The other large 12-pound rim, which hangs horizontally, should be twisted like the first two, but pulled out a little *sidewise* also, so that it can encircle the others easily without touching.

To this horizontal rim, attach three 12″ lengths of gold cord at 10″ intervals around its circumference. Picking up the rim by its strands, and holding it at eye level, adjust the strands until the rim is essentially horizontal. Clip off the ends of the strands to make them even and apply Weldit Cement with a toothpick to their individual tips to keep them from fraying. Set down to dry.

While the strand tips are drying, nip off the tag ends where the strands are attached to the rim, and apply Weldit Cement to the knots with a toothpick. Let dry.

Strike an awl hole near the edge of the disc which will hang in the center (not visible in the photograph). Attach a small fish swivel to it if you plan to use one.

Cut a 20″ length of fine metallic thread and knot it to the swivel (or the disc, as the case may be). Trim off the tag end and apply Weldit Cement to the knot with a toothpick. Let it dry *thoroughly before picking it up* or the knot will pull loose from the weight of the disc.

While all the knots are drying, find the center of balance in the two vertical rims by resting them across the stem of the awl. You may be suprised to find that the center of balance is *not* in what appears to be the center of the curve.)

Strike *tiny* awl holes in the raw edge of the rims at those points, *just* large enough to allow the fine thread to pass through. (Use a stiletto instead of the awl, if you have one.)

HAMOBILE

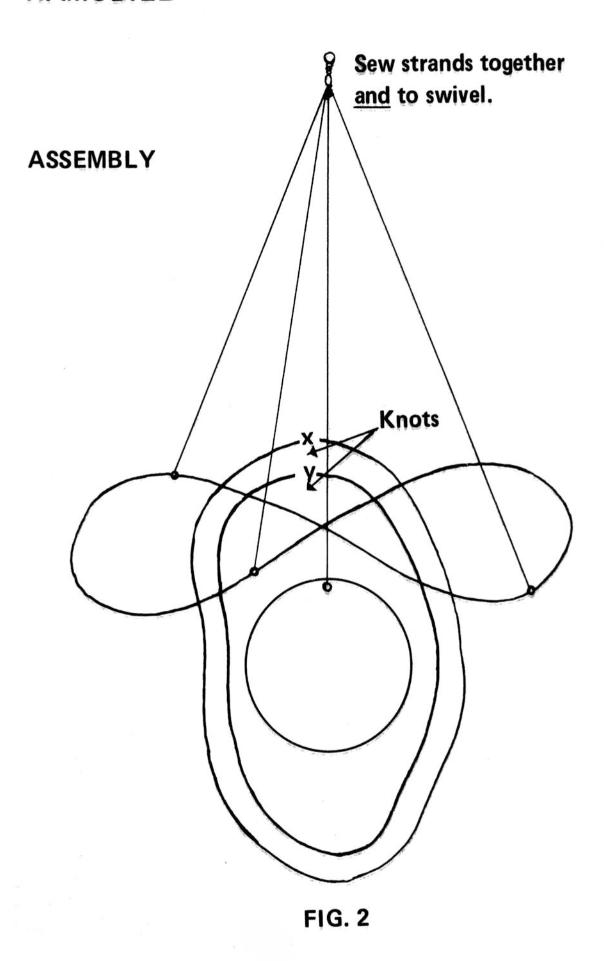

FIG. 2

Make tiny awl holes at balance points, x and y, in raw edge of rims.

Make knots along center strand to keep rims separated.

ASSEMBLE THE MOBILE

There are two ways to assemble the mobile and keep the vertical rims the necessary distance apart: either by passing the thread up through the holes and *tying it around the rims;* or making knots in the thread sufficiently large to *rest the rims on* (Fig. 2). The latter is the better method if you have the patience, because the rims can revolve with greater freedom. But it is hard, with such fine thread, to make the knot large enough. I usually tie the knot three times on top of itself and then apply Weldit Cement to it with a toothpick to build it up even more.

Whichever method you use, the disc should hang in the center of the smaller rim, and the smaller rim should hang between ¼" to ½" below the larger rim.

Slip a needle onto the loose end of the metallic thread; pass it up through the horizontal rim until the vertical rims are nicely surrounded by it; and *sew all the strand tips together and to the swivel or link on the spiral.*

SEQUENCE OF SPIRALS A Suggestion

Consider making a mobile of serpentine spirals in all sizes to float and cast shadows under a white ceiling.

Enchanted by the cut-out animals in the children's book *Noah and His Ark,* Lucy Sargent "brought them all to life" by backing them with tin and bending them into an upright position. With five coats of varnish, the Ark and all its inhabitants will weather any deluge — especially that of eager, playful hands!

The original urn of flowers was painted by a French provincial ceramist in 1750. Unbelievably, you recreate this spectacularly beautiful floral arrangement from a gallon tin can and the pattern on page 59. The fine wires for stems make soldering easy; the grace and harmony of the design make the urn one of the most appealing projects in this book.

Inspiring and atmospheric, inviting meditation, this Mexican *Retablo* was created from a battered commercial fish can and beer-bottle caps. The lunette at the top is a typical Mexican motif, derived from the shell form which the Spanish favored. At the right, the lovely tin Christmas Rose might be considered a worthy offering to the saint depicted in the *retablo*.

DINKYBIRD DINER

This is something you and the children can make on an autumn afternoon.

TOOLS

Snips
Awl
Round-nosed pliers
Long-nosed pliers
Hammer
Sturdy paring knife *(optional)*
Propane torch *(optional)*

MATERIALS

1 3-pound coffee can
1 1-pound coffee can
1 lid from small frozen juice can
Shelf paper, scissors
Rubber cement
20-24 gauge florist wire
¼″ dowel, 10″ long
Solder, Dunton's Tinner's Fluid *(optional)*
Gloves
Wire clothes hanger
Krylon Metal Primer (green), or
Illinois Bronze Tôle Red Primer
Epoxy 220

THE FLOOR

Remove the top rim of the 3-pound coffee can by putting it under the wheel of the can opener *sidewise* (see Removing Rims, page 14).

Remove the *whole end* of the 3-pound coffee can in the same fashion, *or*, by cutting it off with a sturdy paring knife (see Removing Rims).

The *end* of the can, *used under side up*, will be the floor of the feeder. The roof will come out of the sides.

THE ROOF

Wearing gloves, cut through the can along the *left* side of the seam. Pull the can open and step on it. Hammer it well from both sides until it is supple and limber.

Trace off the roof pattern onto shelf paper, marking the spots for the awl holes accurately (Fig. 1). Cut it out and apply it with rubber cement to the *un*lithographed side of the can.

Strike *all* the awl holes *lightly*, right *through* the pattern. Remove pattern and glue.

Restrike the awl holes *along the overlap*, from *both* sides, and hammer them flat.

Strike the two large holes for the dowel (from the *un*lithographed side), forcing the awl through them right up to the handle. Then, since the holes are still not large enough, insert the round-nosed pliers into them, blades closed, and enlarge the holes until the dowel can pass through them easily. Make the holes generous because the dowel will swell in wet weather.

Hammer the roof from the lithographed side, until it curves up nicely.

Cut two 1½″ lengths of florist wire and fold them in half.

Curve the roof around into a cone, so the awl holes in the overlap are centered over each other, and insert the folded wires. Open them up on the inside and flatten them against the roof.

THE WALLS

Remove the bottom lid of the 1-pound coffee can to make an open cylinder.

At one end of the can, cut three small Roman arches, not more than ½″ tall, for the seed to spill out through. One arch should be opposite the seam of the can; the other two on the sides opposite each other (Fig. 5). Your task would be easier were it not for the rim, which is hard to cut through, but which you need for purchase if you intend to glue the feeder together. If soldering, you can remove it.

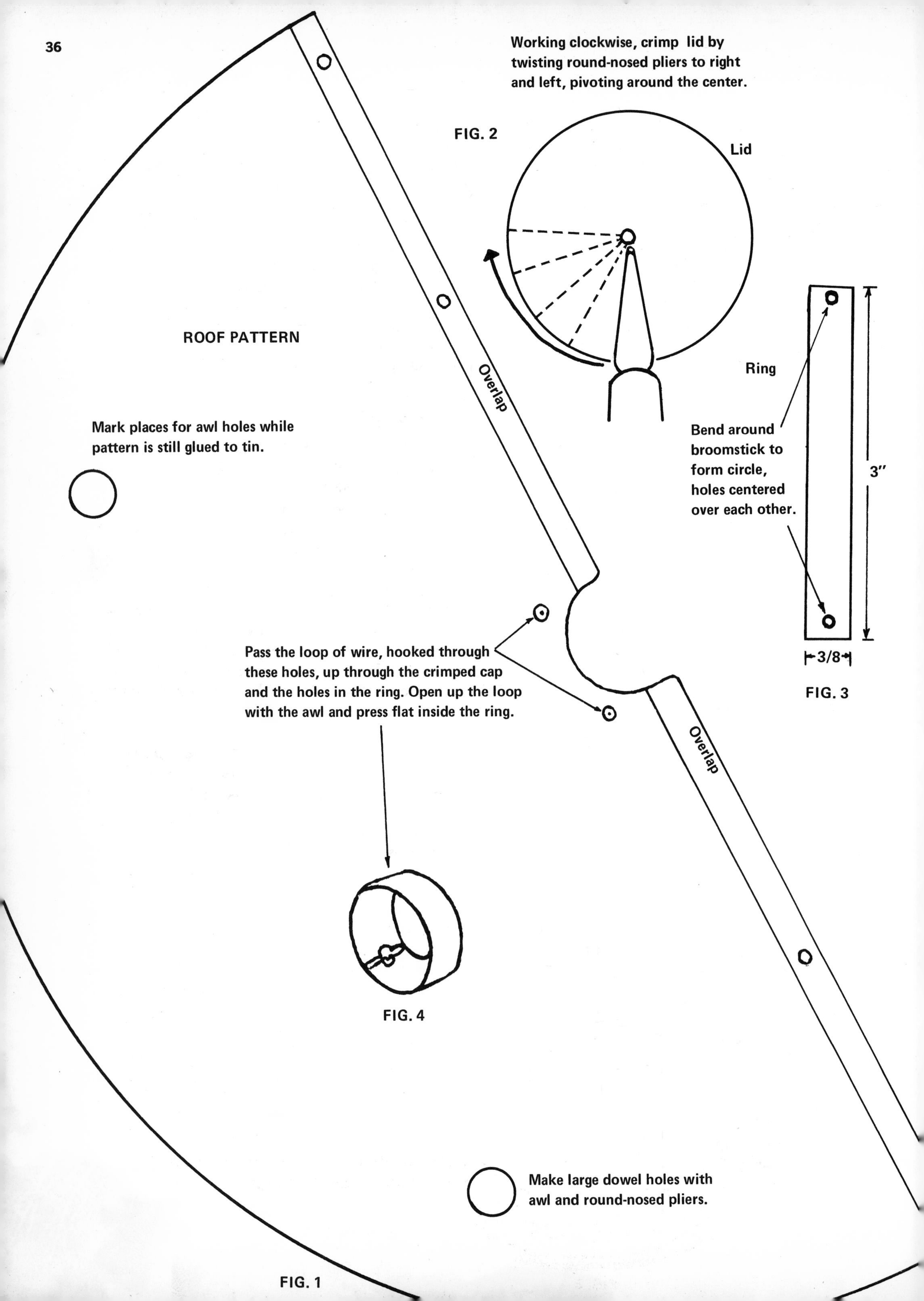
Working clockwise, crimp lid by twisting round-nosed pliers to right and left, pivoting around the center.
FIG. 2
Lid
ROOF PATTERN
Overlap
Ring
Mark places for awl holes while pattern is still glued to tin.
Bend around broomstick to form circle, holes centered over each other.
3"
3/8
FIG. 3
Pass the loop of wire, hooked through these holes, up through the crimped cap and the holes in the ring. Open up the loop with the awl and press flat inside the ring.
Overlap
FIG. 4
Make large dowel holes with awl and round-nosed pliers.
FIG. 1

At the other end of the cylinder, *on the sides, ½″ from the rim,* puncture two holes opposite each other with the awl, plunging it up to the handle. As with the roof, insert the round-nosed pliers and enlarge the holes until the dowel can pass through easily.

Center the walls on the floor and solder or glue the two together. If gluing with Epoxy 220 (which you really need for the sudden strains the bond will be put to when birds light on it), put it in a slow (250°) oven to dry; it should be firm in twenty minutes. Meanwhile, crimp the cap.

THE CAP AND THE RING

Hammer the lid from the small frozen juice can until smooth. Strike an awl hole in the center from both sides and hammer flat.

To crimp the lid, grasp it with the round-nosed pliers so that their tips reach to the awl hole in the center (Fig. 2).

Working clockwise, twist the pliers to right and left, pivoting around the awl hole. Do you see how the tips of the pliers will remain almost stationary while the base travels around the circumference?

Go around again, working only the edge this time, to tighten up the fluting and form a neat little umbrella.

Cut the ring 3″ x ⅜″ from a scrap of tin left from the sides (Fig. 3). Strike the awl holes in each end from *both* sides and hammer flat.

Continue hammering the strip till it curves up; then wrap it around a broomstick until the awl holes are centered over each other.

Cut a 4″ length of florist wire. Poke each end, in turn, through the holes at the top of the roof and twist them tight with the long-nosed pliers.

Squeeze the loop shut and poke it up through the crimped cap and the holes in the ring. Open it up again by pushing the awl through it, and flatten it against the inside of the ring (Fig. 4), with the long-nosed pliers.

THE SKYHOOK

Take the lightest weight wire hanger you can find in the closet and unwind it at the neck with the long-nosed pliers. Straighten the two large curves in it and form a graceful hook on the end.

Spray-paint with Krylon Metal Primer (a lovely leaf green) or Illinois Bronze Tôle Red Primer (a soft rusty red).

As you will see, even a 3-pound coffee can provides such a small platform that only dinky birds can use it. This is an advantage if you have a lot of greedy big birds, but even so, you may want to scale up the whole design by using larger commercial cans. Check out the school cafeteria for these; also restaurants and delicatessens.

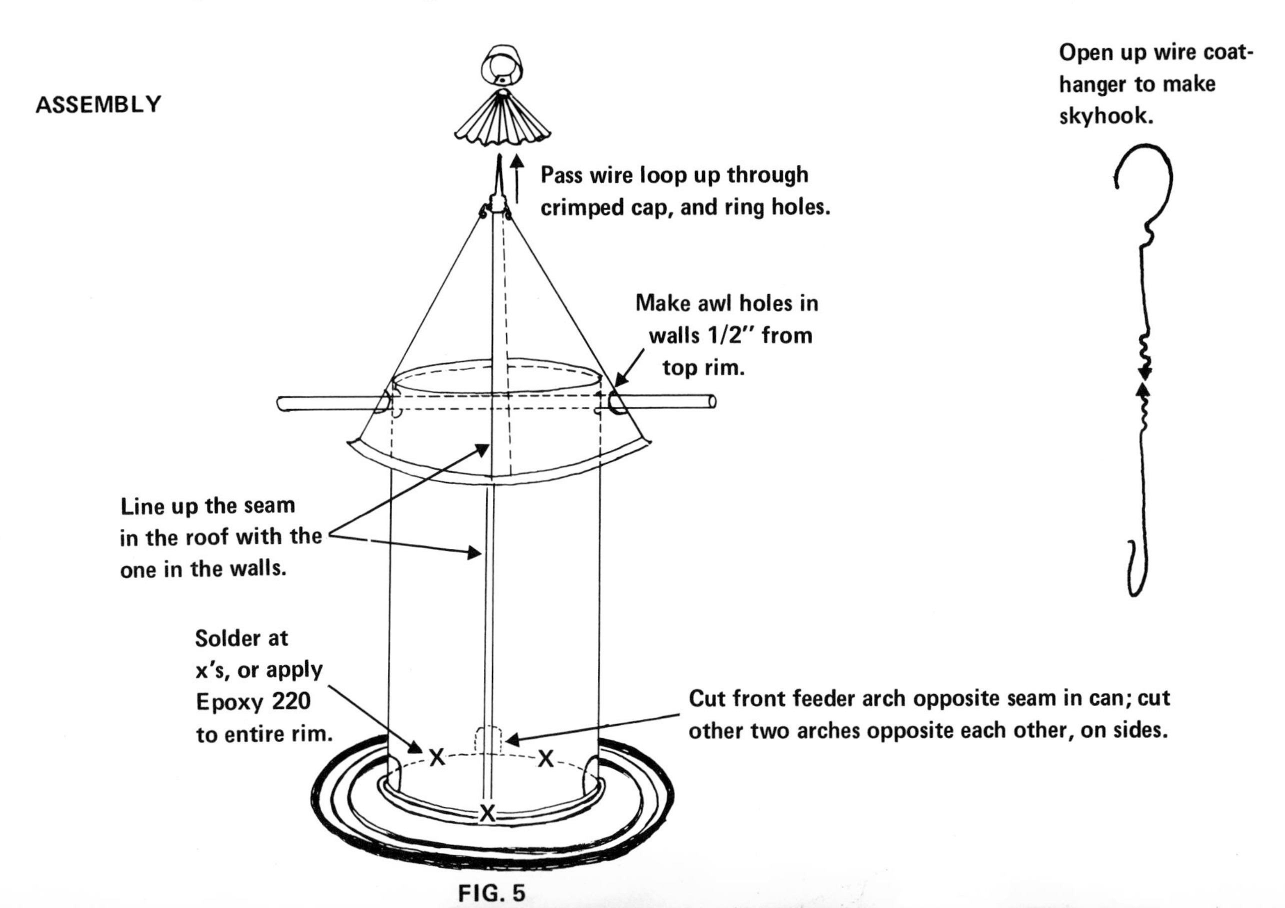

FIG. 5

TWENTIETH-CENTURY PAUL REVERE LANTERN

The Paul Revere Lantern was one design I was determined to make absolutely authentic, and that gave me a good excuse to visit Old Sturbridge Village in Massachusetts, where Mr. Carlton Lees reproduces these lanterns in exactly the same handcrafted manner as the originals fashioned in the early 1800's. The dexterous way in which he runs his heavy, charcoal-heated iron around a lip or up a seam is a beautiful sight to see and one of many reasons for finding an excuse to visit this altogether fascinating restoration.

Unfortunately, however, a Paul Revere lantern is not as easy as Mr. Lees makes it look, requiring special tools, prodigious strength, and a kind of puritanical perseverance which I don't possess. So, sighing, I have come up with a compromise which, authentic though it isn't, looks quaint and casts a lovely light under my gnarled old apple tree.

TOOLS
Plain-edged snips
Hammer
Oyster opener
Awl
Pine board
Long-nosed pliers

MATERIALS
1 3-pound can for the body
1 2-pound can for the cone
1 lid from a small frozen juice can
Architect's Linen, scissors, pencil
Rubber cement
20-24 gauge florist wire
Outdoor light fixture and cord
Illinois Bronze Tôle Red Primer,
or Krylon Flat Black Spray Paint
Gloves

THE BODY

Remove the bottom lid and both rims from the 3-pound can.

Open up the can by cutting along the left side of the seam. Pull it apart *carefully,* just as far as you can without putting a crease in the ribs.

Since the can is the right height but too large around, cut a 4″ strip off *along the seam end.*

On a pine board, with the awl and oyster opener, hammer a decorative design in the ribs *from the inside,* as suggested in Fig. 1.

Strike the awl holes in the overlap from both sides, as always, and hammer them flat. Also strike holes for hanging at *x* and *y* (Fig. 4).

Wearing gloves, reshape the can around a smaller one and wire it together with cotter pins (Fig. 2).

THE CONE

Open up the 2-pound can and hammer it *flat* from both sides. In fact, hammer it on the *un*lithographed side until it curves up nicely.

Trace off the pattern onto the dull side of a piece of Architect's Linen (if you have it, otherwise shelf paper), and apply it to the side of the can with rubber cement (Fig 3). Cut out. Do *not* remove pattern.

With hammer and awl strike in the design *through* the pattern.

Remove the pattern and the glue. Strike the awl holes on the overlap from both sides and hammer them flat.

Again, wearing gloves, carefully curve the cone around until it overlaps, and wire together with cotter pins.

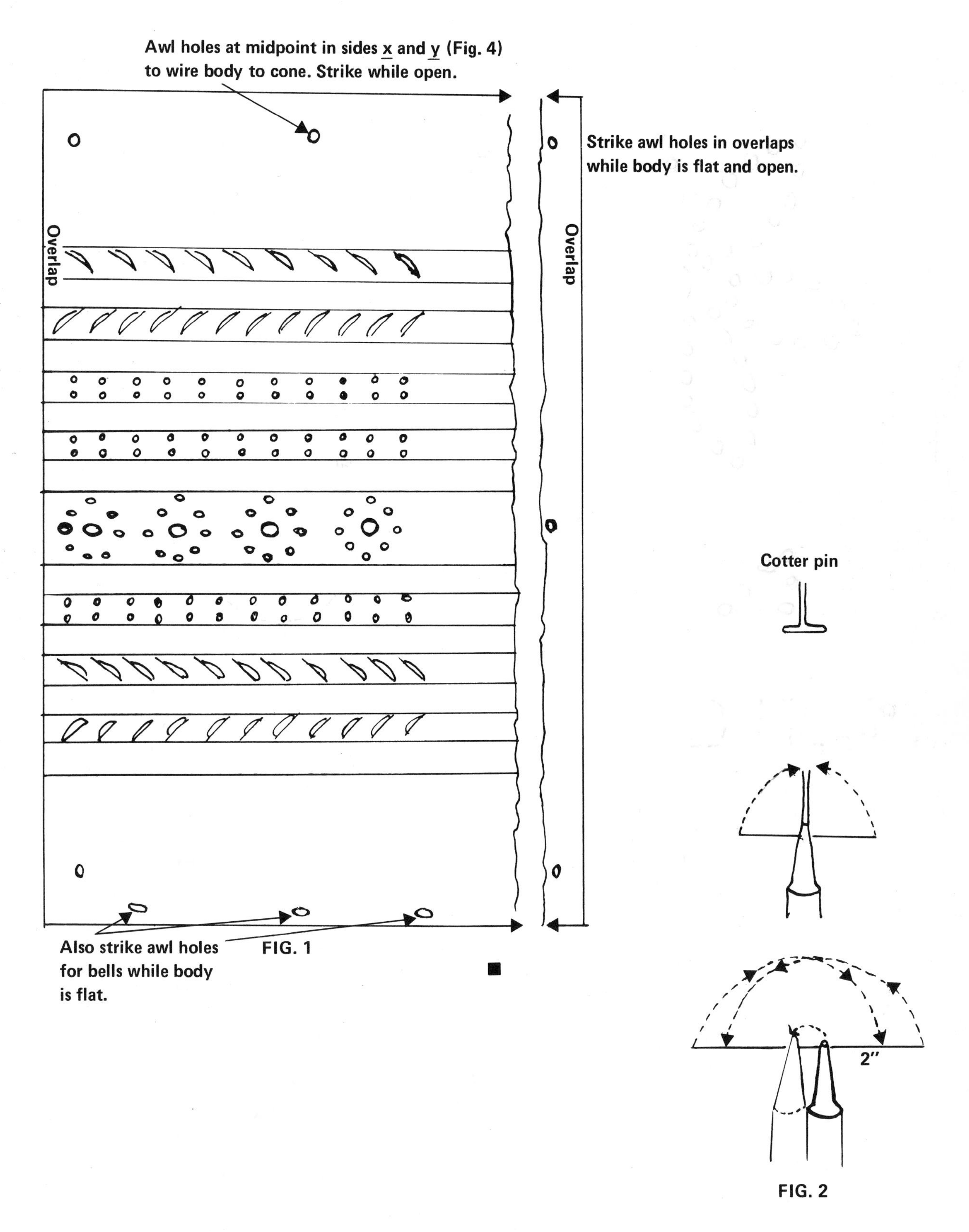
Awl holes at midpoint in sides x and y (Fig. 4) to wire body to cone. Strike while open.
Strike awl holes in overlaps while body is flat and open.
Overlap
Overlap
Also strike awl holes for bells while body is flat.
Cotter pin
2"

FIG. 1

FIG. 2

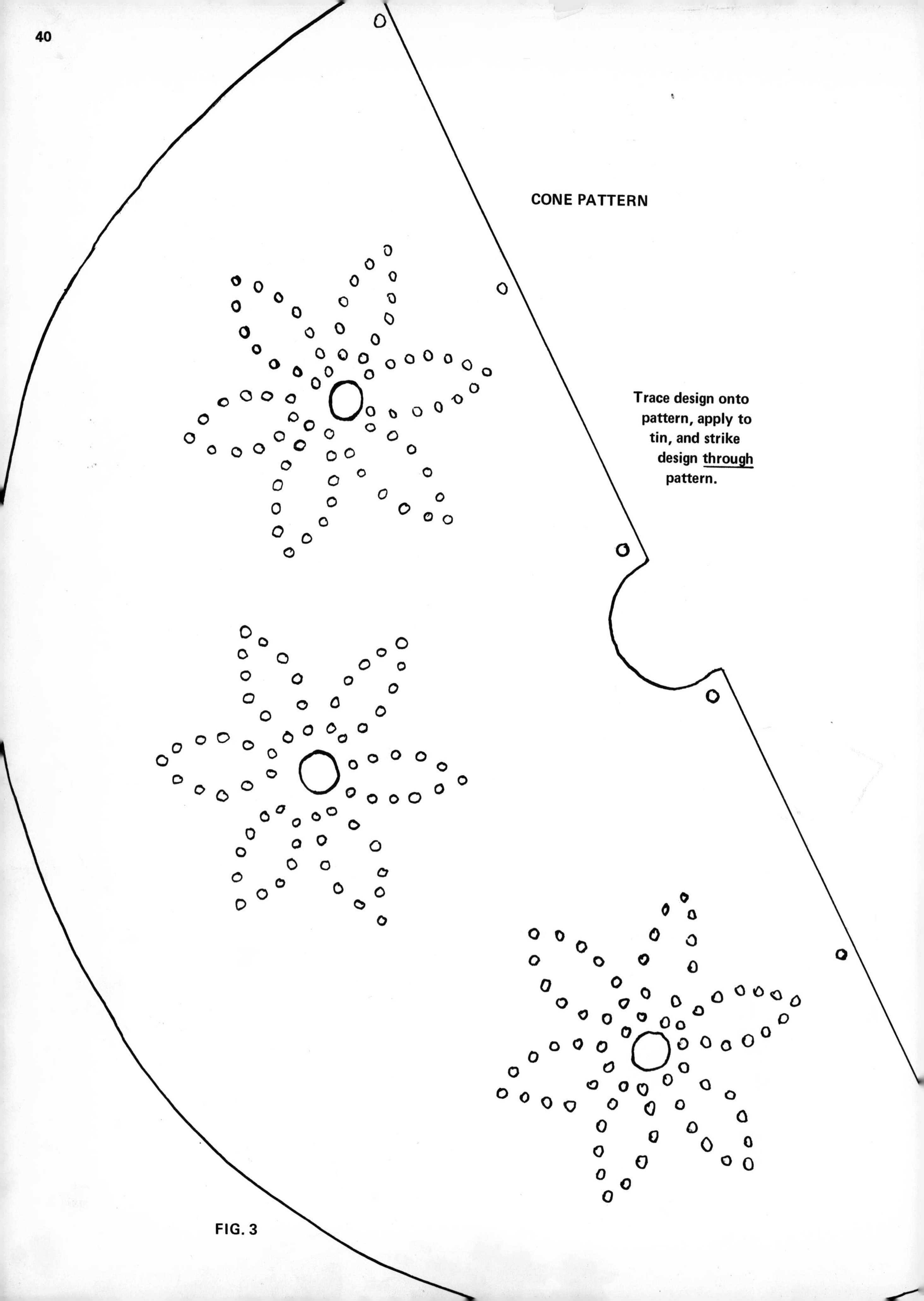

CONE PATTERN

Trace design onto pattern, apply to tin, and strike design <u>through</u> pattern.

FIG. 3

CRIMPING THE CAP AND ASSEMBLING THE LANTERN

In the center of the small frozen juice can lid, make an awl hole large enough for a light cord to pass through it, because this lantern is designed to be electrified. (Consult your local hardware store about the wiring, unless you're a handy electrician yourself. Be sure to use noncorrosive outdoor fixtures, unless, of course, you plan to use it in the house.)

To crimp the cap, refer to the instructions given in the Dinkybird Diner, which just precedes this project.

Before assembling, spray-paint the cone and body *on the outside only,* leaving their interiors silver for better reflection. Spray-paint the crimped cap on *both* sides.

To assemble the lantern, catch 10″ lengths of florist wire in the two awl holes marked *x* and *y* in Fig. 4, and twist them tight with the long-nosed pliers. Send them up through the hole in the cone and hold the lantern by them while you level the cone on the body.

Set the lantern down carefully and *bend the wires* snugly *over* the cone to hold it in position. Feed the wires back inside the cone through a hole in the pierced design.

When wiring the lantern, slip the crimped cap on over the light cord before attaching the plug to the end of it.

ADD BELLS?

Inasmuch as we've abandoned authenticity anyway, we might just as well have some extra fun out of this and add bells to the bottom. The coolest sound you can imagine is made by the simplest bell: a flat, unadorned piece of tin (Fig. 5) cut from the side of a coffee can. You will need eight, spaced about 2″ apart, and hanging every other one at 2″ and 3″. This looks surprisingly Oriental. Spray the bells the same color as the lantern and suspend them with matching wool.

ASSEMBLY

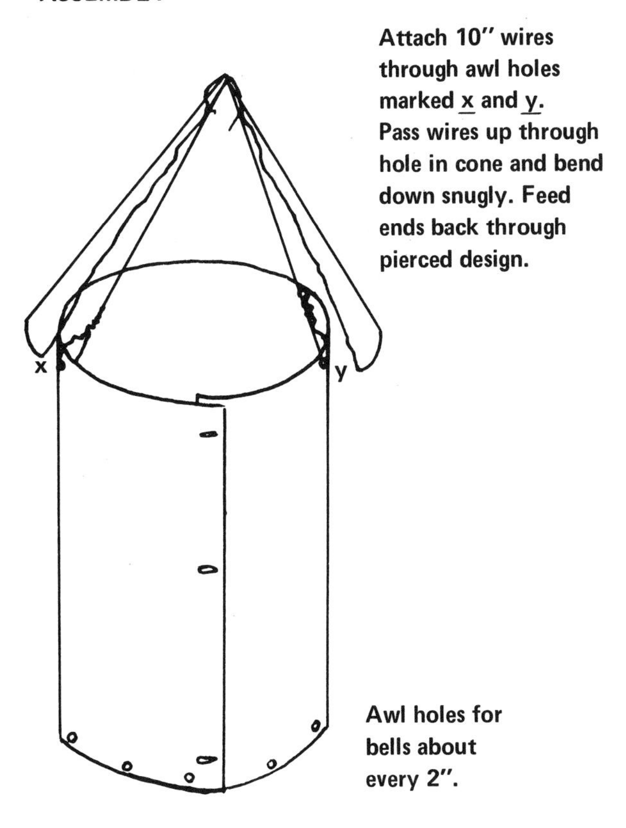

FIG. 4

BELL PATTERN

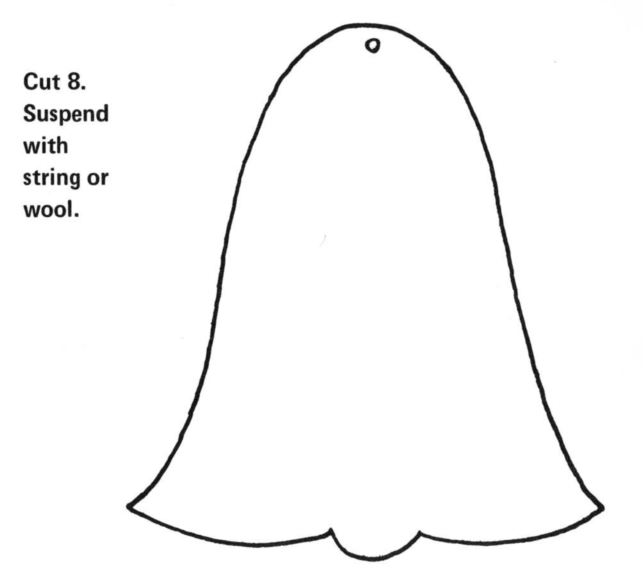

FIG. 5

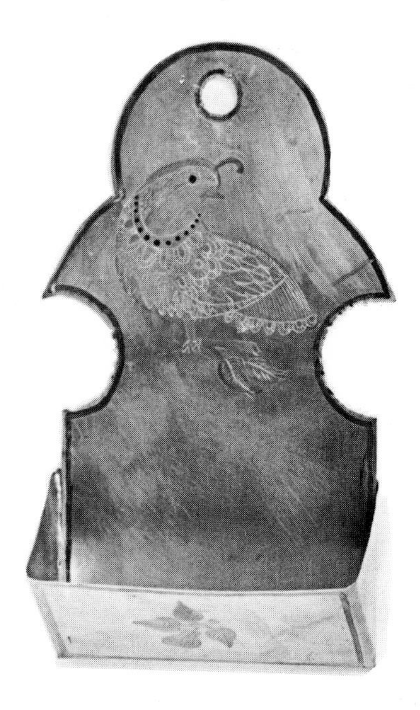

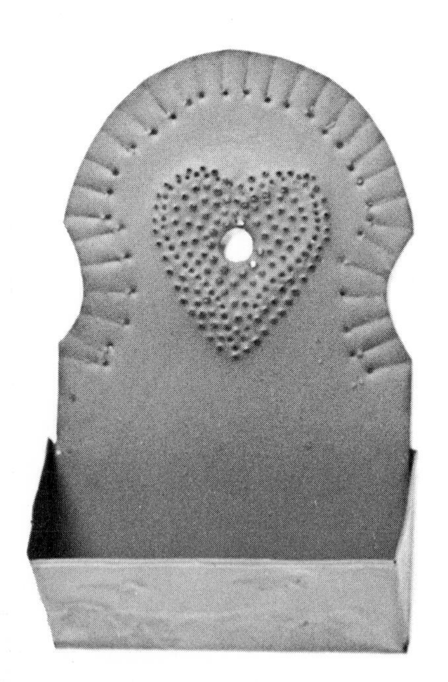

EARLY AMERICAN MATCH HOLDERS

These match holders are more than ornamental accessories; they are really useful articles, and they can be decorated to suit all sorts of situations.

Do sometime try your hand at gilding; it is unbelievably easy and most effective. The partridge was my first attempt, and the wreath on the Easel à la Helen Hobbs the second, so you know it's *got* to be easy. Don't let the list of materials put you off.

TOOLS

Plain-edged snips
Long-nosed pliers
Hammer
Awl
Chisel *(optional)*
Propane torch *(optional)*

MATERIALS

8½" x 10½" piece of tin from 1-gallon can
Architect's Linen, pencil, scissors
Rubber cement
Grease pencil
Epoxy 220 or solder, Dunton's Tinner's Fluid
Carbon paper or lithopone and ballpoint pen *(optional)*
Illinois Bronze Tôle Red Primer

FOR GILDING

Tracing paper, sharpened pencil, turpentine, lithopone
Giles' Black Varnish, Klondike Rich
Gold Bronzing Powder #46, etcher
Plump but pointed brush
#0 Fashion Design brush

NOTE ON GILDING MATERIALS

You can make substitutions for some of these materials: mineral spirits or alcohol can replace the turpentine, chalk for the lithopone, jet dry black enamel for the Giles' Black Varnish (although it is inclined to dry faster than you can work), and a needle stuck in a cork for the etcher. But you will probably have to send away for the bronzing powder and the #0 Fashion Design brush, unless you have a good art store in your community. It's really worth it, though.

BASIC PROCEDURE FOR THE HOLDERS

Decide which of the two match holders you prefer to make (the heart is easier), and whether you want to glue or solder it together. You will note that the tabs must be larger for gluing than for soldering.

Trace the pattern onto a fold of Architect's Linen (see Tracing Patterns, page 16) and glue together while cutting out to get a perfect mirror image. Open up and apply the pattern with rubber cement to the tin; cut out; remove pattern and the glue. Hammer flat all around the edges and improve the outline if necessary.

Unless you have very good (Klenk) snips, or a stamp of some sort, do not attempt to cut the large hole for hanging. Content yourself with a small awl hole, and use a small finishing nail to hang the holder up.

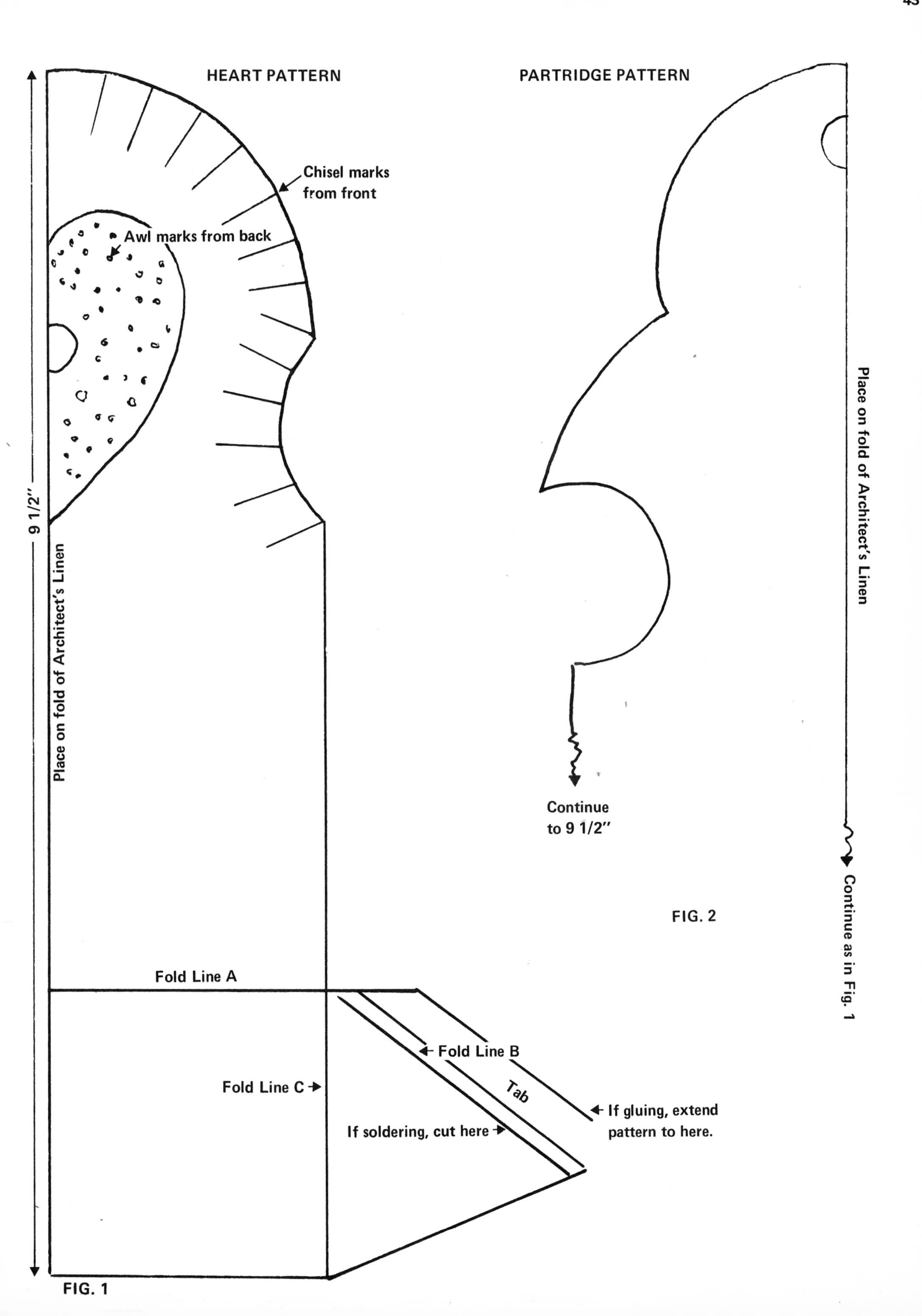
HEART PATTERN
Chisel marks from front
Awl marks from back
9 1/2"
Place on fold of Architect's Linen
Fold Line A
Fold Line B
Tab
Fold Line C
If soldering, cut here
If gluing, extend pattern to here.
PARTRIDGE PATTERN
Place on fold of Architect's Linen
Continue to 9 1/2"
Continue as in Fig. 1

FIG. 1

FIG. 2

If you have decided to make the holder with the heart, trace that form and transfer it to the *back* of the holder with carbon paper (or lithopone) and ball-point pen. Strike in the design with hammer and awl *from the back.* (The heart is more than ornamental; it is a scratcher for ordinary kitchen matches.)

Turn the piece over and draw in the chisel marks around the edge with grease pencil to get them spaced properly before striking them with the hammer and chisel.

Next, score all the fold lines for the pocket *from the front.* Go over them several times until the metal lifts up. This will make it easier for you to bend it with the long-nosed pliers.

If you intend to glue the piece together, now is the time to roughen the tabs and the corresponding places on the back of the holder where the glue will go.

To form the pocket, place the *end* of the ruler on Fold Line A and hold it there firmly as you bend the whole lower section *up.*

Next, with the *blade of the long-nosed pliers running right along Fold Line B,* turn the tabs up at right angles.

Finally, again with the long-nosed pliers held right on Fold Line C, bend up the sides. If you manipulate the long-nosed pliers properly, you will be able to make those sides almost perpendicular to the back of the holder, but still they will have a tendency to spring away from the sides. That's why you need the extra-large tab for purchase when gluing. When soldering, it's neater and nicer to have the narrower flange.

Apply Epoxy 220 to the seams and Scotch-tape in place. Weight it also if you have a metal hammering block, and put it in a slow 250-degree oven.

Gild or spray the holder. The Tôle Red Primer is a beautiful rusty red, very pleasant to live with either indoors or out. For the nifty gilding process, see Bronzing Tin, page 17.

DESIGNS FOR BACK AND POCKET OF PARTRIDGE

EASEL A LA HELEN HOBBS

Helen Hobbs is all things to her students: friend, teacher, mentor, master craftsman, and practitioner-extraordinary of the fine art of tray painting. She ferrets out fabulous designs, lines her pupils up under lamps, and exhorts their every brush stroke. It was her husband, Warfield, who made all our easels; and the minute I saw them, I thought, My, that's for *me*!—meaning *you*, of course!

Easels are easy to make and serviceable, whether for propping up the morning newspaper or something special like an illuminated manuscript.

TOOLS

Snips
Hammer
Awl
Broomstick
Small frozen juice can

MATERIALS

Round 4-gallon commercial can, 10″ x 13″
Paper clips or florist wire
Decora upholsterer's tacks
Materials for gilding *(optional)*

From the flattened sides of a round 4-gallon can (see Tips on Tin Cans, page 9) cut a triangular piece of tin 23″ long, which is 8″ *wide* at one end, *narrowing* to 3″ at the other. Hammer the edges smooth and flat.

Strike two awl holes, *on center,* 5″ from the *wide* end.

Strike another two, *on center,* 3¾″ from the *narrow* end. Hammer them flat (see Striking Awl Holes, page 14).

Place the piece on a counter, face down, with the wide end extending 1½″ over the edge. Carefully, bend it down at a right angle with the heel of your hand.

At a point about 9″ from that first angle, start bending the piece *back* over a small frozen juice can (or similarly sized dowel, etc.) so that it makes the easy curve you see in the photograph.

Curl *up* the narrow end by hammering it around a broomstick or short length of pipe, as you would a scroll (see Scrolls and Candle Cups, page 17).

Since the easel is very limber, it needs to be linked front to back in order to sustain the weight of books. Therefore, nip off the sharp ends of two upholsterer's tacks and bend the stems over to convert them into buttons. Make links of florist wire or paper clips, etcetera, and connect the front panel with the back, hooking the upholsterer's tacks in place as indicated in the diagram.

Don't you want to try gilding? Such fun! See Bronzing Tin, page 17.

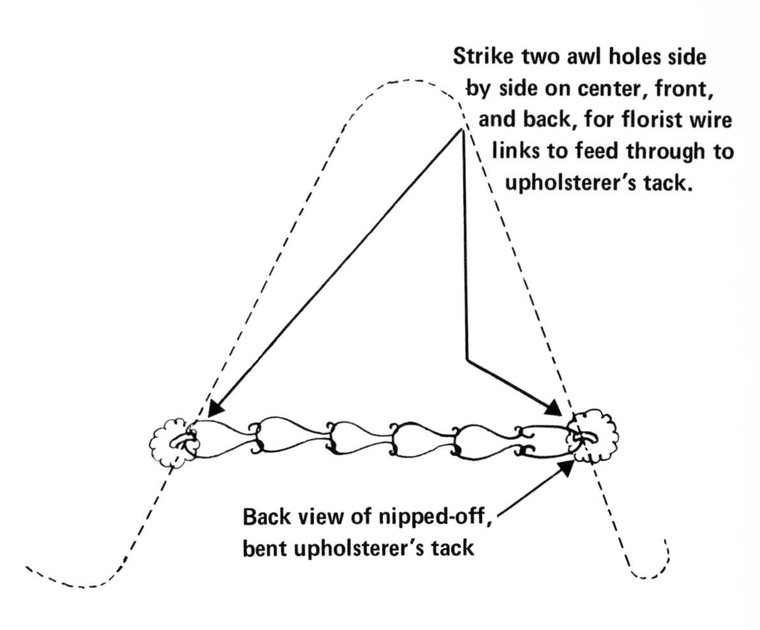

Trace only bold outline onto tin;
paint with black varnish.
Allow to become tacky;
dust with gold powder.
Etch interior lines freehand,
or trace on and etch.

PART THREE
floral fantasies

FLOWERS AND BUTTERFLIES

Flowers are fetching, there's no doubt about it. With the possible exception of the feminine figure, there is no other form in nature more universally appealing. And one of the extraordinary things about flowers—and butterflies, too—is that they don't even have to be naturally represented; they can be frankly fantastic. So, create your own fancies. Make some huge ones for effect, with some tiny ones for contrast; make some cupped, some flared, some tidy, some shaggy.

For gala gatherings it would be foolish to fuss over details. There, bold impressions are important; and it would be wise to double or treble the dimensions suggested here, substituting florist wire, wrapped with floral tape, for soldered stems.

But no matter what the purpose or situation, always use well-hammered lids *so your petals will curve prettily. Those concentric circles can put kinks in your chrysanthemums! Perish forbid!*

INTERLOCUTOR A Construction

Dramatic penetration of perimeters, flying off from a florescent focal point, tin fronds reflecting the fire of the flaming arrow.

APOTHECARY ROSE

The very essence of a rose, and easy, easy to make.

TOOLS
Plain-edged snips
Ball-pein hammer
Propane torch *(optional)*
Vise *(optional)*
Asbestos mat *(optional)*

MATERIALS
2 soup can lids per rose
Sash cord wire, florist wire, or
16-gauge wire, depending upon use
Architect's Linen, scissors, pencil
Rubber cement
Weldit Cement or solder, Dunton's Tinner's Fluid
Beer bottle cap
Green Floratape *(optional)*
Glitter and glue *(optional)*

CUT THE PETAL LAYERS

Select the choicest gold-lined (matching) lids you can find. Remove the price mark, if any, and hammer flat.

Divide each lid into halves, cutting down to within a pennyworth of the center (Fig. 1).

Divide those halves into *thirds,* making six petals.

Round each section as shown in the diagram, cutting all six sections in one direction before turning the piece over to round the sections in the other direction. This will help you cut smoother curves faster. (You *could* put ripples in the petals if you wanted to bother, but even without, they will look quite rose-y.)

Hammer each individual petal, gold side up, in a beer bottle cap, using the round end of the ball-pein hammer. Don't be fussy about it; you want a few crinkles in the edges.

WIRE THE PETAL LAYERS TOGETHER

There are three different ways to wire and assemble petal layers, depending upon the purpose the flowers will be put to.

If you want to make a quick, casual bouquet, strike *two* awl holes in the center of the petal layers and send ordinary florist wire up and down through them, twisting it around itself tightly at the back of the rose and winding it with green Floratape. Daub the center of the rose with glitter glue and glitter, to conceal the wire. Because of the tight wiring, this method requires no glue between the layers.

If you want them to last forever and be truly authentic tôle roses, they should be soldered onto a length of 16-gauge wire that projects through a *single* hole in the center of the petal layers like a pistil, with a little drop of solder or gesso on the tip of it.

If you want something a little more realistic and very pretty (see the Nosegay for a Tenth Wedding Anniversary, page 160), strike *one* awl hole in the center *large* enough to allow *sash cord wire,* cut to desired length, to pass through it. Fray one end of the wire to a depth of ⅜″ to resemble stamens. Score the petals around the awl hole. Apply Weldit Cement to the petals *and* to the sash wire at the base of the stamens; allow to become tacky; then send the wire through the petal and pull snug. Add more Weldit if it seems advisable. (Even better, solder the rose together from the back, held head down in the vise, so the solder flows down along the stem through the petal layers.)

CUT THE LEAVES

Here again you have a choice, depending upon the durability required. The quicker, but less durable, method would be to trace off the *whole* leaf pattern (onto Architect's Linen preferably, since you will need more than one) and cut it out, *in one piece,* from the side of a soup can, twisting the stem around the rose stem, and covering the joint with Floratape. (See alternatives presented in Fig. 2.)

The authentic way, shown in the diagram, involves cutting separate leaves (as plain or realistic as you wish) and soldering them onto bent stems made of either 20- or 24-gauge florist wire. Someday I hope you will discover how easy—and satisfying—this process can be (see Soldering, page 20).

These roses are charming just as they are, gold and silver, but there will be times when you will want to paint them. Tôle roses were always matte in finish, and satin varnish with artists' oils will give you that, and still be fairly quick-drying. Spray paint is the easiest and quickest, but must be done before assembling. It might be fun to two-tone them, like hybrid teas.

PETAL DIAGRAM

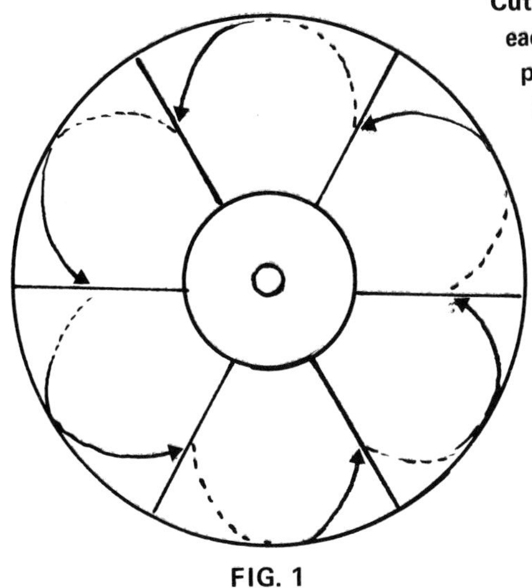

FIG. 1

Cut two layers for each rose; alternate petals when assembling.

Divide lid of soup can into six sections. Round each section in one direction; turn over and round in the other.
Beat each petal from top with round end of ball-pein hammer in beer bottle cap.

LEAF DIAGRAM

If gluing, cut whole spray from side of can. Wrap stem around wire and bind with Floratape.

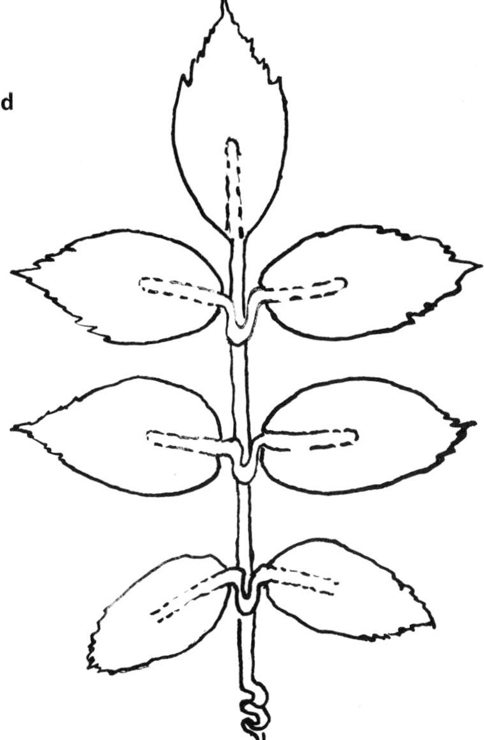

FIG. 2

If soldering, cut 7 separate leaves. Solder them in pairs onto bent stems, as shown. Solder these pairs in turn onto main stem already topped by end leaf.

CHRISTMAS ROSE

Helleborus niger, while not a true rose, has many of the characteristics and all of the charm of one; and because it blooms for weeks in winter, it is called the Christmas rose. Though the blossom is simple enough, the leaf requires a little fussing, but isn't it fabulous? (See color photograph, page 34.)

TOOLS

Plain-edged snips
Awl
Screen installation tool *(optional)*
Ball-pein hammer and board
Round-nosed pliers

MATERIALS

3-inch lid
2-inch lid
Side of large can, preferably
unribbed and unlithographed
Architect's Linen, scissors, pencil
Rubber cement
Sash cord wire for stems to desired length
Weldit Cement or Epoxy 220
Krylon Flat White Spray Enamel
Krylon Metal Primer
Tube of yellow acrylic paint

THE LEAF

If you plan to make a centerpiece requiring a number of Christmas Roses, by all means use Architect's Linen for the leaf pattern. Trace only the bold outlines, omitting the serrations. Cut it out and apply with rubber cement to the side of the can.

Cut the leaf out, hammer it flat, and spray-paint it lightly with Krylon Metal Primer.

Draw in the serrations with pencil if you feel you need to; otherwise, just cut them freehand, looking to the book for guidance.

Hammer the leaf flat again and vein it as heavily as possible, either with the awl *from the back,* or with the screen installation tool from *both front and back* (see Veining, page 16).

With the round-nosed pliers, *curve* all the outer edges *under,* just the slightest, softest bit.

Unravel a length of sash cord wire to a depth of 1½", into the six separate strands of which it is made. Glue these individually along the veins in the lobes of the leaf with Weldit Cement or Epoxy 220. Let dry.

Spray-paint with Krylon Metal Primer.

THE BLOSSOM

Here again use Architect's Linen for the pattern if you plan to make a number of blossoms.

After cutting out the petal layers, hammer them flat and improve their silhouettes. Hammer again.

Strike awl holes in their centers large enough to admit the sash cord wire.

Vein each petal with the awl or screen installation tool.

Cup the blossom by beating it *from the front* with the *round* end of the ball-pein hammer on a *flat* board. This is a very simple, open blossom, with none of the recurved edges found in many roses.

Fray the end of a length of sash cord wire to a depth of ½", this time separating every filament of wire in the six strands.

Slip the blossom on and secure it with Weldit Cement or Epoxy 220.

Spray-paint the blossom on both sides with Krylon Flat White Spray Enamel. When dry, put a paper bonnet over the blossom and spray the stem with Krylon Metal Primer.

THE SEPAL

The sepal pattern used here is a craftsman's reluctant compromise with Nature.

Make the pattern as usual; cut out the sepals; and hammer them until they curve up.

Spray-paint them on both sides with Krylon Metal Primer.

When dry, slip them on behind the petals so they curve *back,* and fix them in place with Weldit or epoxy.

As a final touch, daub the tip of each individual stamen in the blossom with yellow acrylic paint, thick from the tube.

Isn't it innocent?

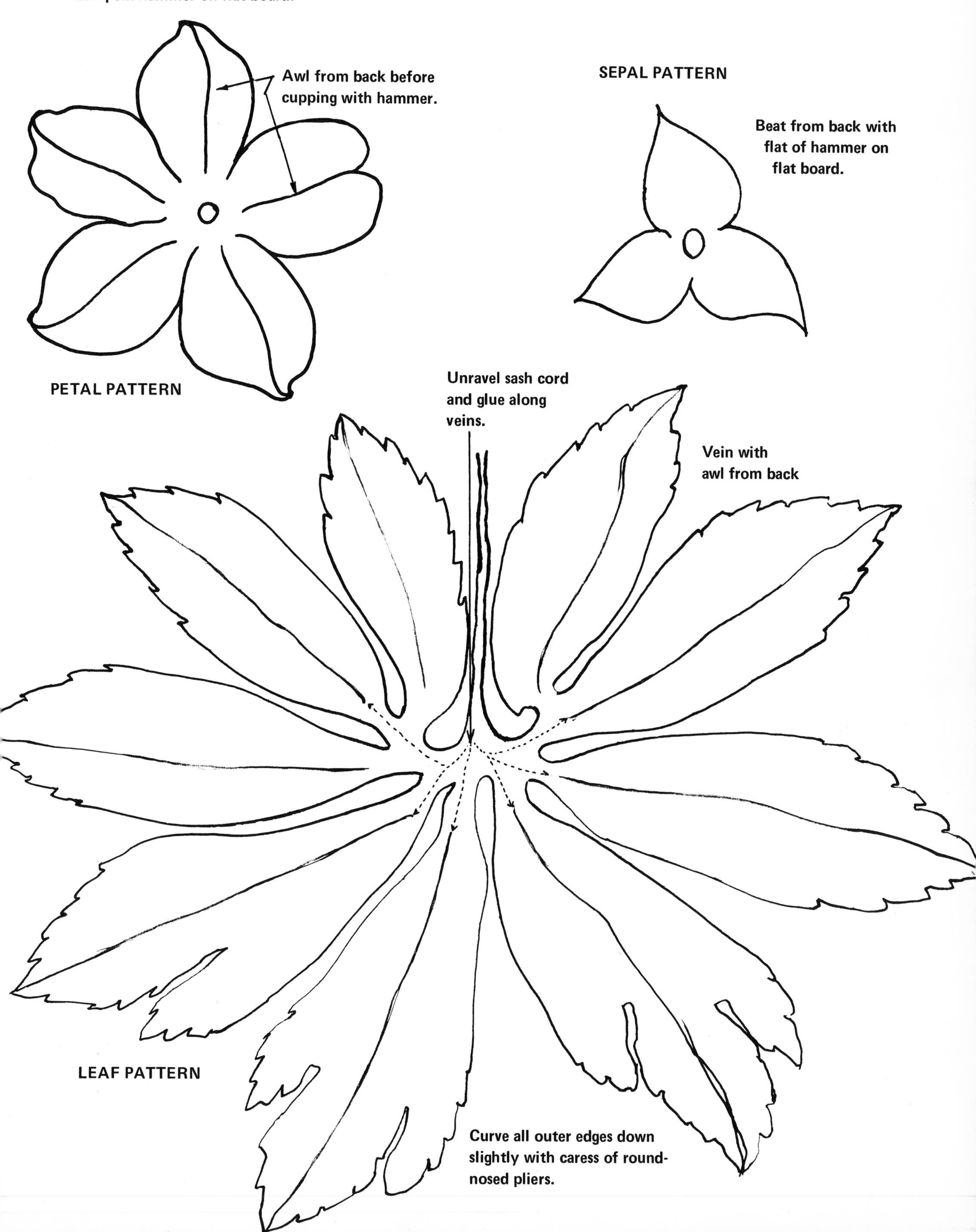
Beat from front with round end of ball-pein hammer on flat board.
Awl from back before cupping with hammer.
SEPAL PATTERN
Beat from back with flat of hammer on flat board.
PETAL PATTERN
Unravel sash cord and glue along veins.
Vein with awl from back
LEAF PATTERN
Curve all outer edges down slightly with caress of round-nosed pliers.

FRENCH TOLE FILLER FLOWERS

No period arrangement would be complete without its complement of little filler flowers, those smaller blooms that fill in a mass bouquet, serving as foils for the extravagant rose-lily-chrysanthemum forms. Here are two no-nothing filler fellows taken from an eighteenth-century French collection. (See color photograph, page 68.)

TOOLS	MATERIALS
Plain-edged snips	Lids from small frozen juice cans
Awl	Sides of finely ribbed cans
Ball-pein hammer and board	Bottle caps
Round-nosed pliers	Solder or Epoxy 220 *(optional)*
Propane torch *(optional)*	16-gauge wire, or florist wire and Floratape
	Paints

As with most flowers, you have a choice as to how you will assemble them. The originals of these filler flowers were soldered onto a length of 16-gauge wire that projected through their centers like a pistil, with a drop of solder congealed on the tip to resemble a stigma. It might be that you could substitute Epoxy 220 for solder, but with a *single* set of petals, I'm not sure that there would be enough purchase to make a solid bond. As an alternative, however, you can wire the flower together, striking *two* awl holes in the center and sending lightweight florist wire up through and down again, twisting it around itself tightly and winding it with Floratape.

THE FLOWERS

Cut down the lids from the small frozen juice cans to 1½″ in diameter.

Strike one or two awl holes in their centers (depending upon the method of assembly) from both sides, and hammer flat.

Divide the lids in halves, then in quarters. Hammer flat.

Cut the petals as indicated in the diagrams and shape them with the ball-pein hammer in a bottle cap.

THE LEAVES

Here again you have a choice as to whether you will cut *long stems* on the leaves to wind around the flower stem and bind with Floratape, or whether you will solder on *stemless* leaves.

As you can see from the diagram, I cut leaves, freehand, from the side of a finely ribbed can, because it so closely resembled the hand-ribbed original, but do whatever pleases you.

PAINT THE FLOWERS

There are options here also. The originals were done in artists' oils (mixed with satin varnish). Dri-Mark permanent pens are the easiest, but look less realistic and substantial. Talens Glass Paints are rich and very durable, but transparent, and therefore not truly authentic. In the end, as always, it's a matter of personal preference and consideration of the purpose.

REGAL LILY

This is an average-size, naturalistic lily. (See color photograph on page 68.)

TOOLS	MATERIALS
Plain-edged snips	Sides of *un*ribbed cans
Ball-pein hammer and board	Ruler, felt-tip pen, turpentine
Screen installation tool or awl	Weldit Cement
Round-nosed pliers	16-gauge wire
Long-nosed pliers	Krylon Flat White Spray Paint
	Krylon Metal Primer *(optional)*
	Tube of yellow acrylic paint

From the side of an unribbed can, measure off and cut a rectangular piece of tin, 3¼″ x 5½″, using ruler and felt-tip pen.

Rule the piece off, as shown in the diagram, into five 1″ sections, plus a ½″ section for stamens, and a ¼″ border at the bottom.

With felt-tip pen, draw in the wavery petal design freehand, and cut out as directed.

Hammer flat, clean up any rough edges, and remove pen marks with turpentine.

Rule in the veins, preferably with the screen installation tool (see Veining, page 16), or with awl.

Hammer the petals on either side of the vein on a board, from the *back*, with the *round* end of the ball-pein hammer.

Smooth Weldit Cement along the border *under the petals* (*not* under the stamens). Allow to become tacky.

Using the round-nosed pliers, *roll up* the lily *tightly*, and hold it closed with the long-nosed pliers while you pull out the petals and stamens to form the flower. (By the time you have finished fussing, the rolled border should be dry.)

Insert wire stem into plier hole in roll and paint as you wish.

FRENCH TOLE FILLER FLOWERS

PLUMBAGO

Beat in beer bottle cap from top with round end of ball-pein hammer.

Snick edges under with round-nosed pliers.

POTENTILLA

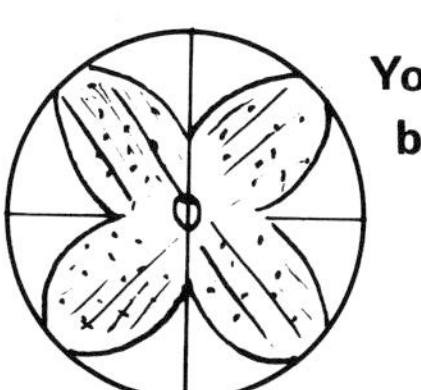

You can texture a blossom by coating with gesso and combing.

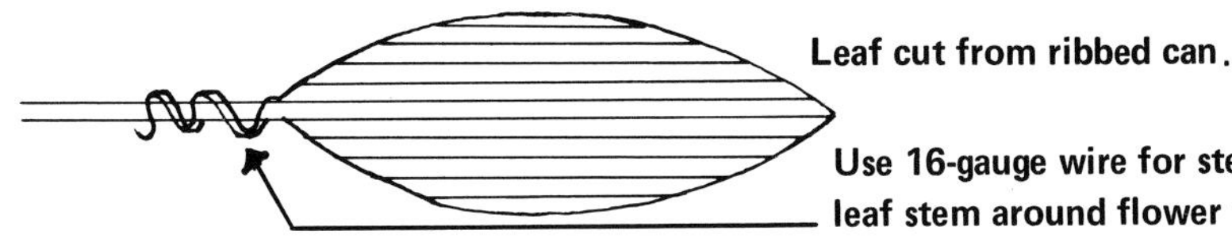

REGAL LILY

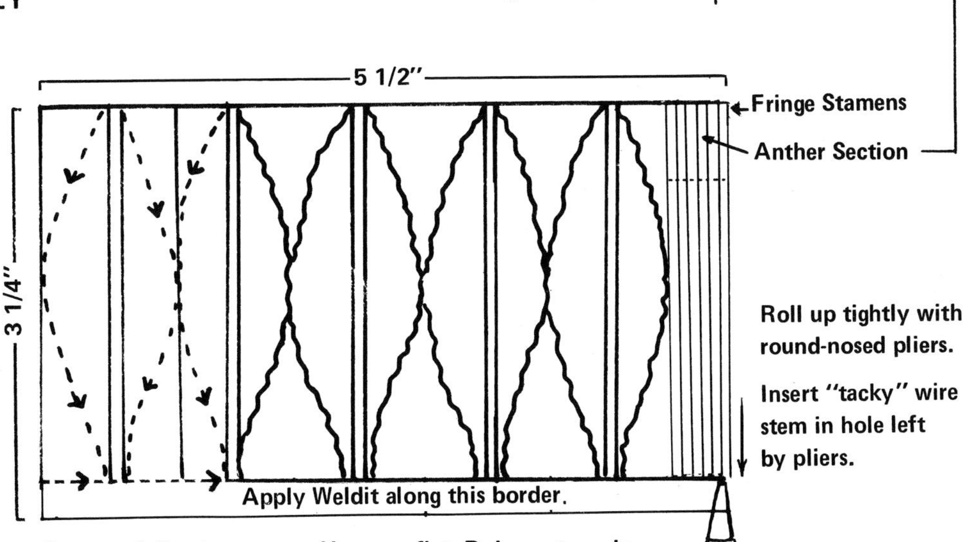

Cut out, following arrows. Hammer flat. Rule center vein, preferably with screen installation tool: convex end from top; concave from back. Beat with round end of ball-pein hammer, from back, on either side of vein, on board.

CRAZY CHRYSANTHEMUM

This hybrid wonder is made on the same principle as the Regal Lily, that is, it's *rolled up.*

Unless you plan to paint it, cut with serrated snips from a shiny piece of tin so it sparkles.

Hammer it *perfectly flat* before rolling it. (If you're painting it, do so while it's flat.)

Apply Weldit as indicated in the diagram; roll with round-nosed pliers; hold with long-nosed pliers and *pull out and spiral every strand of fringe* to form a ravishing swirl.

(This flower adorns The Hat, page 159.)

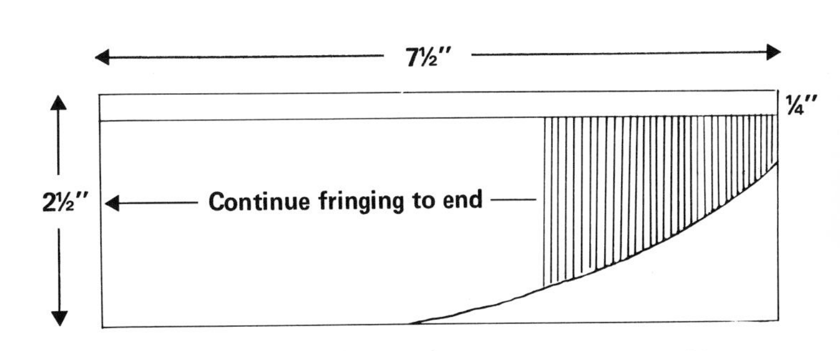

Cut curve and fringe, holding fingers of helping hand in the way to prevent curling. Hammer flat from inside, working back and forth across the strip from the base of fringe to tips. Roll up tightly with round-nosed pliers from narrow end. Pinch each strand between thumb and third finger to curve. Spiral tip of each strand slightly.

URN PATTERN

With awl from top, vein as much or as little as you wish.

x

y

x

y

y

y

x

x

Paint all x's same color for unity.

x

x

All y's can be extra pieces glued or soldered on.

x

FRENCH URN OF FLOWERS

Painted in 1750 by a French provincial ceramist, there is nothing provincial or even French about this urn of flowers (see color photograph, page 34) except as it reflects the Cantonese vogue popular in France at the time. It is a highly cosmopolitan collection which includes tobacco, potato, and pepper plants from the New World, as well as lotus and chrysanthemums from the Far East. The design lends itself handsomely to sculpture, and if you have not already learned to solder, let this be the piece to inspire you—the fine wires make it so easy. If that thought daunts you, try using just the central urn section as a sconce or perhaps a bookend.

TOOLS

Plain-edged snips
Awl
Ball-pein hammer and board
Toothpaste cap
Propane torch
Asbestos mat

MATERIALS

1-gallon can
Sides of *un*lacquered cans
Architect's Linen, scissors, pencil
Rubber cement
Tracing paper and lithopone
Solder, Dunton's Tinner's Fluid
Artists' oils and brush
Weldit Cement

THE URN

Trace off the urn pattern, cut out, and apply with rubber cement to the flattened side of a 1-gallon can (olive oil, gasoline, etcetera, having removed the lithographing, of course).

Cut out with plain-edged snips. (See Cutting Out Patterned Designs, page 16)

Remove the pattern and the glue; hammer flat.

Improve the outline until it's clean and smooth. Again hammer flat.

With the awl from the front, *vein* the leaves as much as you wish, and *outline* the petals (see Veining, page 16).

Give dimension to the fruits and flowers by beating them with the *round* end of the ball-pein hammer from the *back* on a board. Try to work *within* the lines you have just scored. Rescore them from the front, if necessary, when you've finished hammering.

Having shaped the flowers, you will want to shape the urn. Trace the water-lily and volute design onto the urn, using lithopone (or chalk) on the back of your tracing paper (see Tracing Patterns, page 16).

Score in the design with the awl *from the front.* Then raise all the areas marked *x* by beating them with the *round* end of the ball-pein hammer from the *back.* Try not to hit the scored lines with the hammer. Rescore them from the top if necessary.

On the lip of the urn, I emphasized those crescent markings by hammering across it from the back with the top of the toothpaste cap held at an angle. It's not really necessary to do this; you can always paint in highlights and shadows if you'd rather. Or perhaps you'd like the lip just as well plain. Suit yourself.

If you are making the entire sculpture, wait to add the flower forms marked *y* until you have cut, and soldered on, all the wired branches. If, however, you like the urn without those branches (it would make a stunning sconce just as it is), then proceed to cut and glue on the *y* forms. They, too, have to be beaten, or fringed, depending on which forms you decide to use. I put chrysanthemums on both sides, but that is perhaps too obviously symmetrical, and you might prefer to substitute a passion flower for one of them.

THE BRANCHES

If you have never soldered before, I would suggest that you cut all the flowers and leaves from the side of an *un*lacquered can. Perhaps the gallon can you started with is unlacquered, in which case you can safely continue with it. But lacquer burns when heat is applied to it too long. Sometimes this can make for an interesting gradation in color, shading from near-black to clear brown to amber to gold. When you become experienced, you can use this technique intentionally and most effectively. But if you are just learning, it is easier to start with plain tin (see Soldering, page 20).

When cutting a lot of anything, always use Architect's Linen; otherwise, with paper the patterns tear, and you have to make new ones. Of course, cutting anything as small as leaves and buds really calls for those beautiful Klenk snips, but then, what doesn't? They make all the difference between work and play. Treat yourself.

Do all veining of leaves and beating of petals *before* you solder their wires on. Notice that most wires are soldered on the back of the flowers, but that there are twelve (flanking the urn) which are soldered from the *front* to form plump ovaries.

Notice also that those little *half-opened* tobacco

blossoms are soldered onto *the same wire* as the *fully-opened* blossom, rather than being wired separately. So also with a few leaves that have no wire at all. Add more leaves if you like, but remember, in assembling the various pieces, to *keep the arrangement tight,* laying it on the diagram as you add a unit to be sure you've maintained the scale.

Because no book could accommodate the whole design intact, you will have to reverse the assembly procedure for the opposite side of the urn, and you will have to study the photograph on page 34 to see what elements comprise that top-center spray.

Once you have assembled and soldered everything in place, including all the branches *and the urn-backing,* then it is time to add the *y* flower forms on the front. The chrysanthemum is made from two discs (cut-down lids from small frozen juice cans or whatever), and the passion flower could be either a fringed disc or a fringed bottle cap, depending upon how elegantly or primitively you are approaching the piece.

As for the lotus blossom, the seed section goes on first (did you make it convincingly bumpy?) and then the petal section. If necessary, score that horizontal line running between the petals *once more* from the top and bend the petals up (or down) to increase the sense of dimension. Glue in place with Weldit Cement.

Now you are ready to paint—and that is always a matter of personal taste. You will notice that my colors are low-keyed and Oriental in feeling, reminiscent of the jade and rose quartz trees made in China. To achieve that intriguing pearlescence, pour a half teaspoon of spar varnish into a bottle cap and blend in the tiniest snick of oil paint from the tube with your brush. Lots of varnish not only ensures transparency, but also quick drying, which is a real blessing. It would take days otherwise. Furthermore, if you plan to leave any areas of the tin *un*painted, as I did, you will find the high gloss of the varnish mixture most compatible with it—far better than a flat paint finish. The Talens Transparent Glass Paints would be excellent for this also. But whatever paint you use, make all the *x* areas the same color to unify the composition and keep the center of gravity low.

Wait till your friends catch sight of this one!

FLOWER PATTERNS

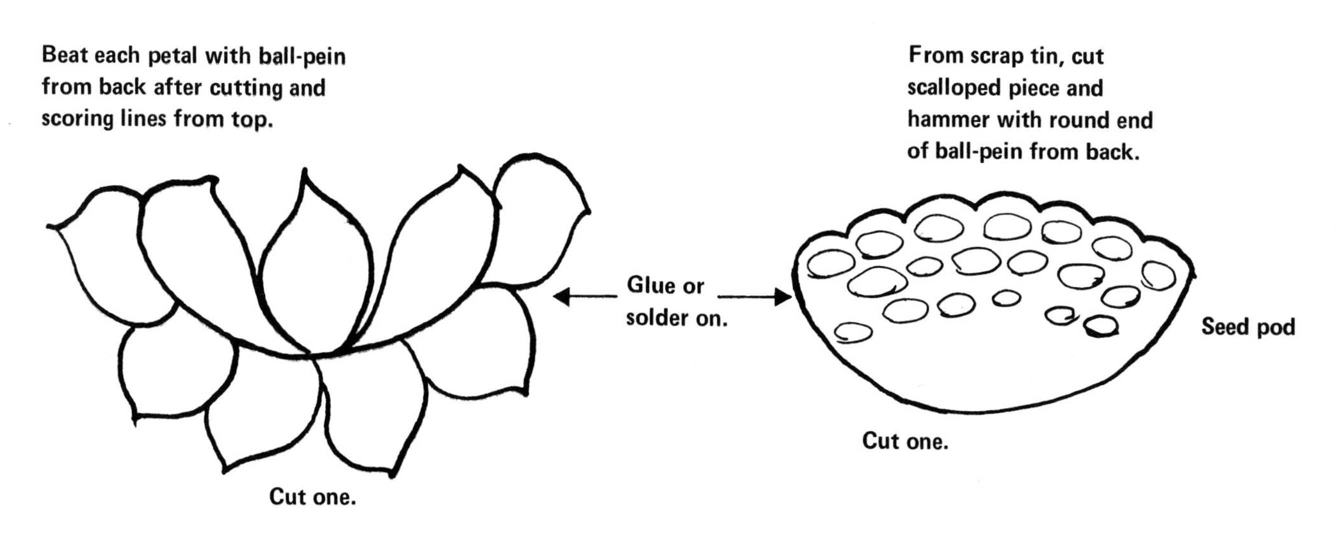

Lotus Blossom

PLACEMENT OF WIRED FLOWERS

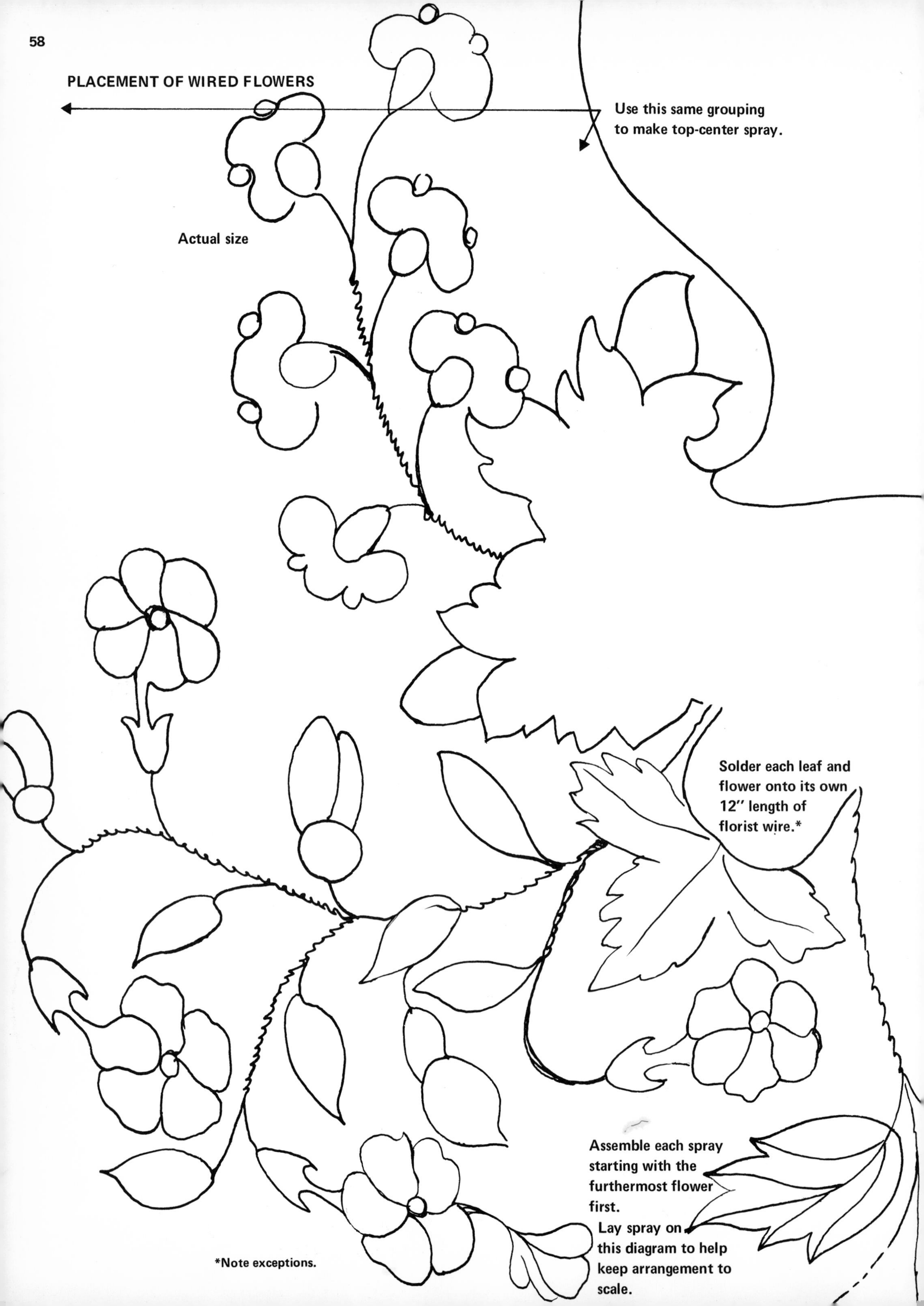

Use this same grouping to make top-center spray.

Actual size

Solder each leaf and flower onto its own 12" length of florist wire.*

Assemble each spray starting with the furthermost flower first.

Lay spray on this diagram to help keep arrangement to scale.

*Note exceptions.

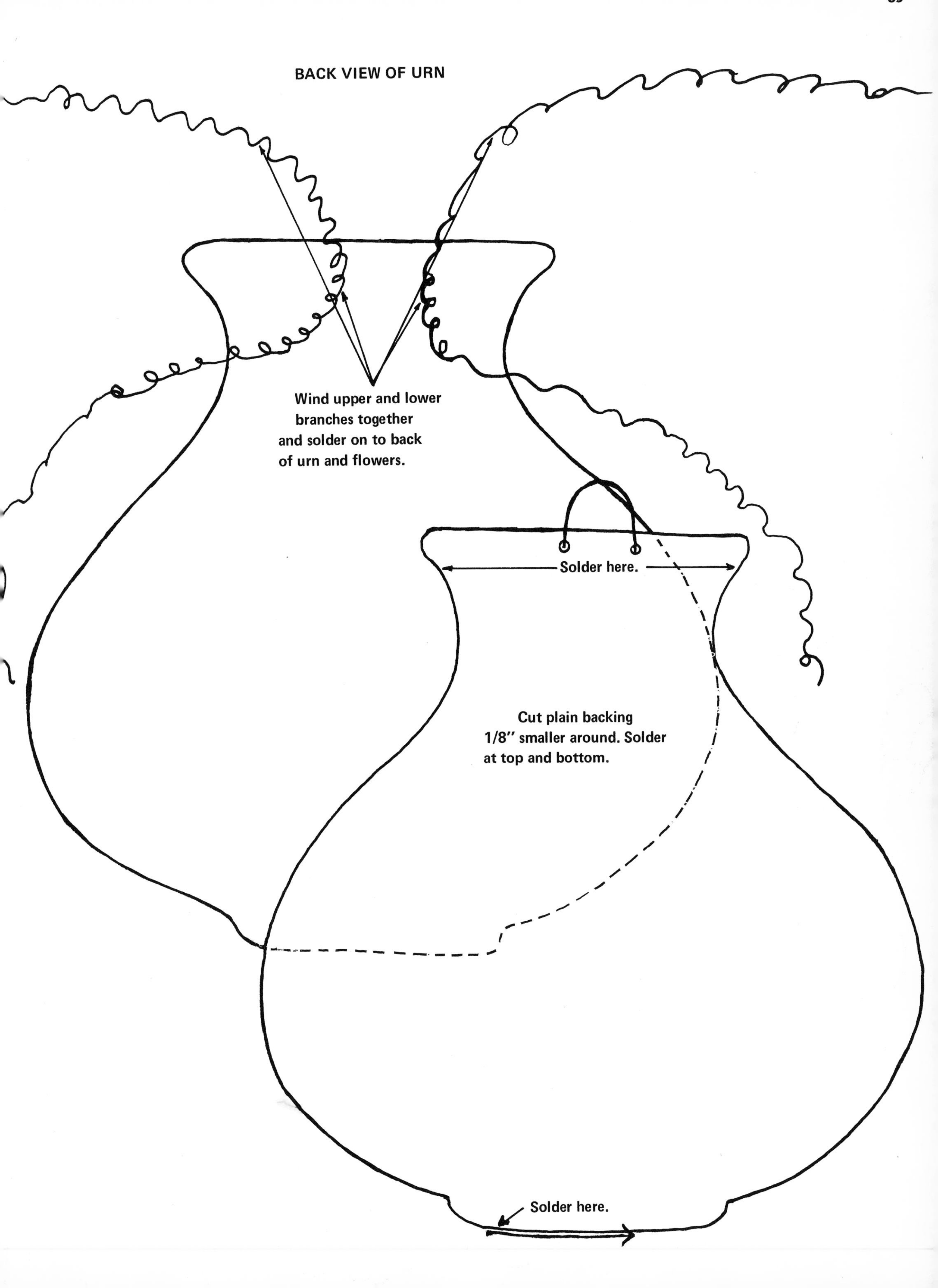
BACK VIEW OF URN
Wind upper and lower
branches together
and solder on to back
of urn and flowers.
Solder here.
Cut plain backing
1/8" smaller around. Solder
at top and bottom.
Solder here.

BUTTERFLIES

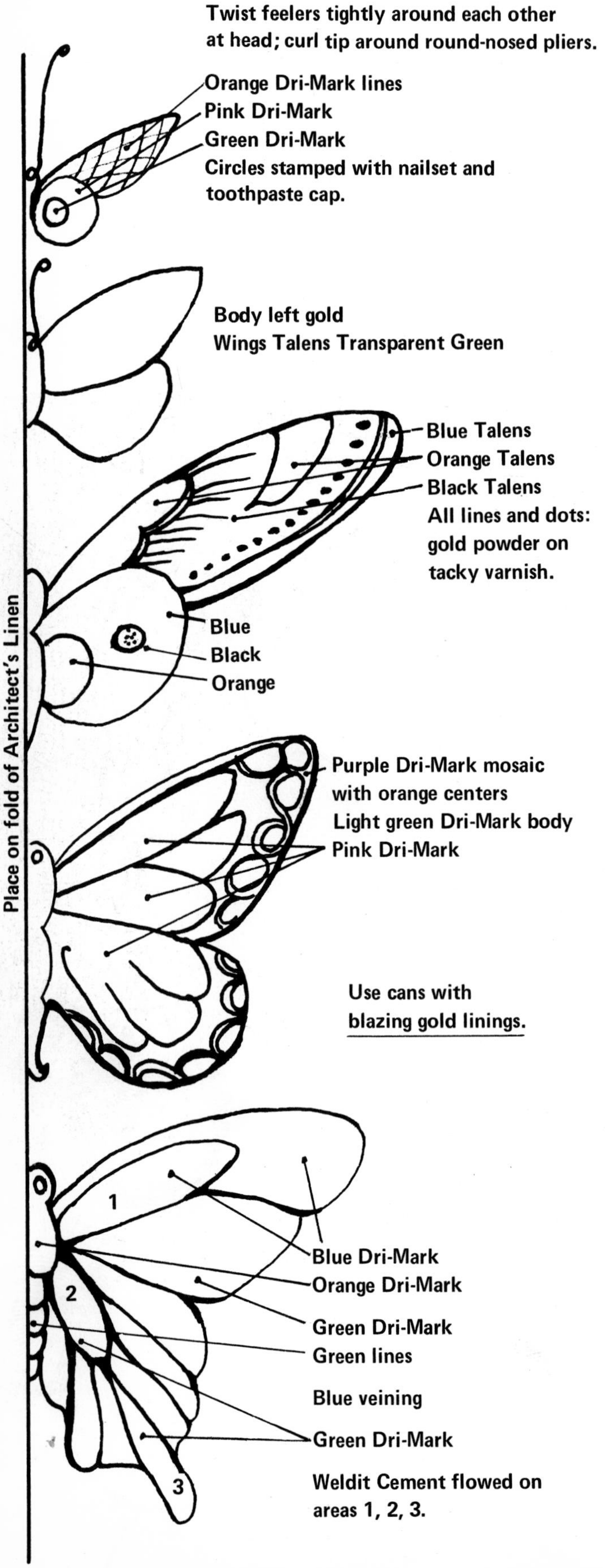

Twist feelers tightly around each other at head; curl tip around round-nosed pliers.
Orange Dri-Mark lines
Pink Dri-Mark
Green Dri-Mark
Circles stamped with nailset and toothpaste cap.

Body left gold
Wings Talens Transparent Green

Blue Talens
Orange Talens
Black Talens
All lines and dots: gold powder on tacky varnish.

Blue
Black
Orange

Place on fold of Architect's Linen

Purple Dri-Mark mosaic with orange centers
Light green Dri-Mark body
Pink Dri-Mark

Use cans with blazing gold linings.

1
2
3
Blue Dri-Mark
Orange Dri-Mark
Green Dri-Mark
Green lines
Blue veining
Green Dri-Mark
Weldit Cement flowed on areas 1, 2, 3.

If you love to fuss, this is for you! And they make *such* a nice gift. (See color photograph, page 101)

TOOLS
Snips
Hammer and board
Awl
Round-nosed pliers
Screwdriver
Nailset and toothpaste cap
Metal hammering block

MATERIALS
Sides of cans with *blazing gold linings*
Architect's Linen, scissors, pencil
Rubber cement
26-gauge brass wire for feelers
20-24 gauge florist wire
Epoxy 220

SUGGESTED PAINTS
Dri-Mark *fineline* pens
Talens Transparent Glass Paints
Giles' Black Varnish
Baer's Bronzing Powder: Klondike Rich Gold #46
#0 Fashion Design brush
Weldit Cement, used decoratively

GENERAL PROCEDURE

Trace off pattern on a fold of Architect's Linen and *glue the fold together* with rubber cement *before* cutting it out to ensure a perfect mirror image on the wings.

Open and apply the pattern with rubber cement to the lacquered side of the can and *carefully* cut *out.* (You really need good snips for this fine work.) Hammer flat.

Strike two tiny awl holes in the head from the *back,* through which to thread the fine brass wire for feelers.

If you plan to wire the butterfly onto something, strike two more small awl holes midway on either side of the body, this time from the *top.* Use the same brass wire here also.

With hammer and screwdriver from the top, *outline* the body and the line between the upper and lower sets of wings.

With hammer and screwdriver from the *back, ridge* the body *crosswise.*

Paint as you please. Even a plain emerald-green one is lovely!

If mounting these butterflies on long limber florist wires to tuck into a flower arrangement, hammer the wires *flat on one end* on a metal hammering block. Glue with Epoxy 220 onto the back of the body.

COMMENTS

Even though the outlines of these butterflies have purposely been kept simple, they are sensitive. They must flow with perfect grace, no corners. And since color is the eye-catcher, it should be intense and clear. That's why I suggest the Dri-Mark fineline permanent pens and the Talens Transparent Glass Paints.

Applying paint to the *lacquered* side of the butterfly will increase the color richness. But also look upon the lacquer as one of your colors. In other words, paint in the design in such a way as to leave gold borders around them—around the edge of the wings, too.

Note that Weldit Cement can be used decoratively, flowing it on over key areas to give dimension and texture, not to mention added brilliance to the color under it.

Try the butterfly that is pink and green. It's lovely to live with.

ORIENTAL POPPY An Idea

It is part of woman's creative instinct to put familiar things in a new context. Good flower arrangers do this continually. Here Mrs. Alden Vose responded intuitively to a need in her new herb garden by bringing out tiebacks from the house to crown the gateposts.

GARDEN LABELS

Most garden labels have an unfortunate way of attracting attention to themselves and littering up an otherwise serene and gracious planting. My herb garden, which is fairly formal and filled with unfamiliar specimens, called for quantities of labels that were readable but retiring, rigid enough to stand up neatly, and rugged enough to withstand the rigors of winter. The classic wreath I came up with has proven satisfactory in every way, and I am sorry not to have a photograph which really shows it to advantage. Here it is angled, ready to be placed in the ground, and you can only surmise how sweetly the wreath surrounds the written word. The other shapes were experiments which did not suit my situation, but might well suit others. They are easier to cut out, and when you need a good many, that's a definite advantage. Best of all, they may give you ideas.

TOOLS

Serrated snips
Hammer and board
Awl

MATERIALS

Sides of sturdy, unribbed cans
Krylon Metal Primer

I suggest using serrated snips simply because the leaves on the wreath are made with four little clicks of the serrations. The other designs would be really better made with plain-edged snips.

Even though you will presumably be making a lot of labels, you don't need to use Architect's Linen for the pattern. Simply use the first label you make as a blank for the others.

An easy way to make an oval is to remove the bottom of a small frozen juice can and squeeze the sides with your hand until you have the shape you like. Then trace around it.

You may find you like to use the perfectly plain oval as your garden label (Fig. 1), and not bother with the wreath. But if you do find the wreath appealing, perhaps you should make a template out of cardboard (Fig. 2), to indicate where to make the *initial* cuts for the leaves. I am too impatient to bother with that step. I just use my eye and listen to the clicks (Fig. 3). After you've made a few, you'll get into the rhythm of it.

Krylon Metal Primer is designed expressly to provide a protective coating for metal in the environment, and being olive green, it also harmonizes happily with nature.

I use an awl to scratch in the name of the plant. You may prefer to use a white china marker, available at stationer's stores. After you've written on your label, bend it over a counter edge to angle it for easy reading.

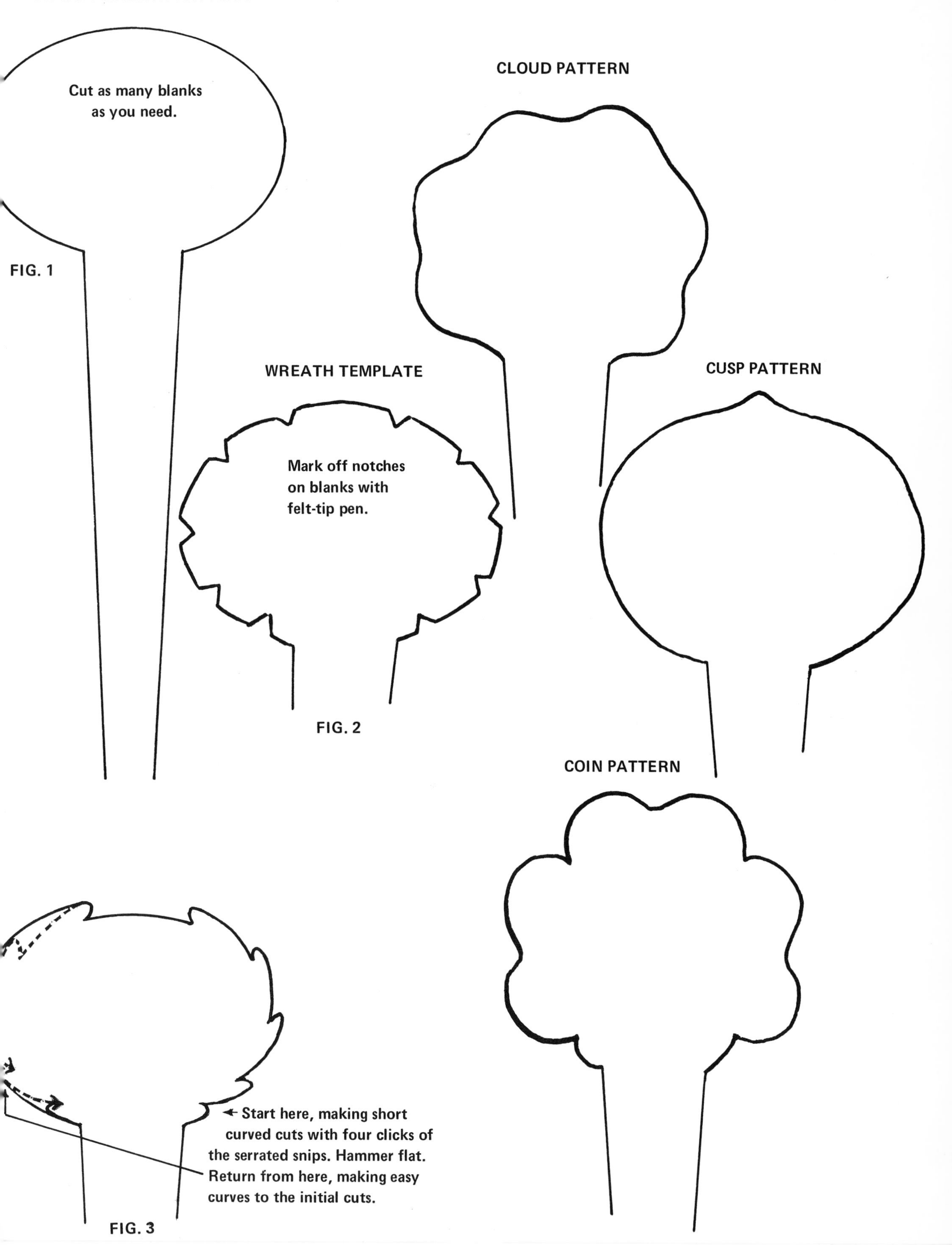
CLASSIC WREATH PATTERN
Cut as many blanks
as you need.
FIG. 1
CLOUD PATTERN
WREATH TEMPLATE
Mark off notches
on blanks with
felt-tip pen.
FIG. 2
CUSP PATTERN
COIN PATTERN
Start here, making short
curved cuts with four clicks of
the serrated snips. Hammer flat.
Return from here, making easy
curves to the initial cuts.
FIG. 3

CATALYST An Assemblage

Though predominantly monochromatic because of weathered wood and dried materials, chiaroscuro is provided by the shining tin leaves; leaves, which in spite of their metallic crispness, undulate in the same sinuous fashion as the natural striations in the wood and fruiting scape. The circular rose forms, centered on the diagonals of the frond, control the fluidity of the other materials. Several coats of satin varnish cast a unifying sheen over the whole composition.

Imagine a friend's amusement to find this lovely May Basket, filled with flowers, on her doorstep!

CATCHY CONTAINERS

People who have made a study of the subject say that the consummation of a craft occurs when a skillfully executed design combines function with flair. I can only think that that *must account for the satisfaction one gets in creating a catchy container. Especially when one picks a few posies to put in it and presents it to a friend for the pure pleasure it gives them both.*

MAY BASKET

Glue or solder it, whichever you prefer, but remember to cut an *odd* number of ribs.

TOOLS

Plain-edged snips
Hammer and board
Long-nosed pliers
Propane torch *(optional)*
Asbestos mat *(optional)*

MATERIALS

1 lid from a 2-pound coffee can
20-24 gauge florist wire
Weldit Cement or solder, Dunton's Tinner's Fluid
1 25¢ piece
Felt-tip pen and turpentine

CUT THE RIBS

Hammer the lid until smooth.

Place a quarter in the center and trace around it with a felt-tip pen.

Divide the lid into eight equal sections (see Dividing Lids Into Equal Sections, page 15), cutting down *to* the center circle.

Divide seven of those eight sections in *halves.*

Divide the eighth section in *thirds!* (That's the way you get an odd number of ribs.)

Continue to divide each section in *halves* until you reach the one you divided in thirds. Stop there. You should have 31 sections.

Hammer the lid flat, working from the center out.

Pare down each section, as shown in the diagram, to form *straight ribs.* Remove felt-tip pen marks with turpentine.

WEAVE THE BASKET

Hook the florist wire around the base of a rib, and squeeze it tight with the long-nosed pliers.

Weave under and over the ribs, pulling the wire sufficiently tight to form a nice little nest.

Bind the wire around the rim of the basket by bending the tip of each rib down over it.

BRAID THE HANDLE

Cut three 12" lengths of florist wire. Twist them together at one end with the pliers and tack them to a board.

Braid them together.

Nip off both ends neatly and feed them down through the openwork on opposite sides of the basket.

Secure with solder or glue at the base of the ribs.

The bow is extra, and you can see that it is just a twist of braided wire. In the color photograph on page 67, a little yellow bird is perching there between the loops.

METHOD OF OBTAINING AN ODD NUMBER OF RIBS

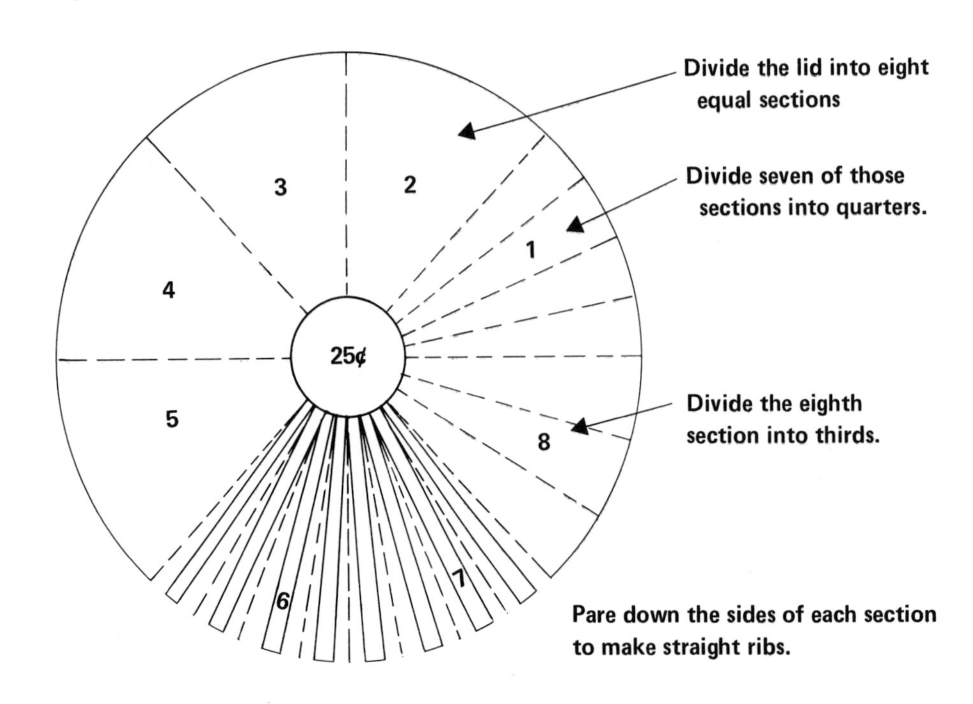

The May Basket, filled with lovely spring flowers, hangs over four variations of the Japanese Lansket. Used as either basket or lantern, the lansket has limitless possibilities for use indoors or out. The Bonbon Basket in the foreground once held sardines; now it becomes an elegant serving accessory.

The French tôle filler flowers are invaluable blooms that round out a mass bouquet. Here they are displayed with the Regal Lily in their own simple arrangement, showing how versatile these little flowers can be. Clockwise from the top: two kinds of "Love," the Little Pink Pig, and the Heart Keepsake — whimsical valentine greetings for special friends.

BONBON BASKET

A confection would be cordially received at Christmastime or any other if presented in a sprightly container such as this. And who would ever dream it had once held sardines! (See color photograph, page 67)

TOOLS

Snips
Long-nosed pliers
Hammer and board

MATERIALS

Portuguese sardine can
1 yard of ⅜″ ribbon
Shelf paper
Rubber cement
Weldit Cement

CUT THE RIBS

Cut a strip of shelf paper 1″ wide to fit exactly around the outside of a Portuguese sardine can.

In order to obtain an *odd* number of ribs (which you *must* have for weaving a basket), fold the paper *almost* in half but not quite, making one side ⅜″ longer than the other (see diagram).

Continue to double the paper over *up to that point*, creasing the folds well, until all the sections are ⅜″ wide.

Paint the side of the can with rubber cement and smooth the creased paper around it.

Cut through the rim down along the creases to within ⅛″ from the bottom of the can.

WEAVE THE BASKET

Cement one end of the ribbon to the base of a rib, and weave in and out, pushing the sections forward and back to prevent cutting the ribbon. (If you plan to use the container for something other than food, spray-paint it before weaving, and that will dull those sharp edges on the sections.)

When you reach the top, clip off the ribbon, cement the end to a rib, and wire in a bow at each end.

Cut strip of paper to fit around can.

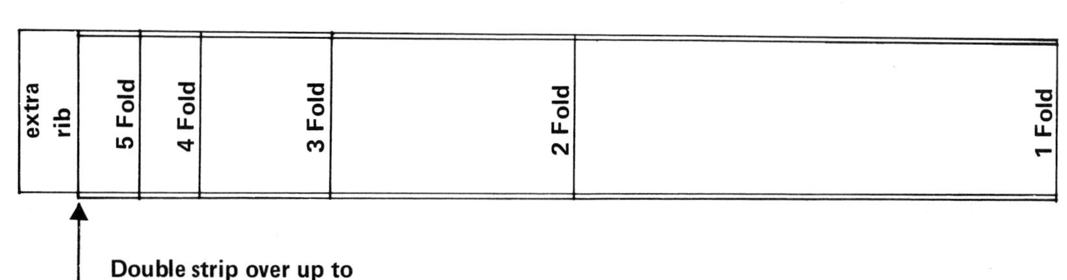

Double strip over up to here. Continue to double it over to that point four more times. Crease sharply. Unfold and apply to can side coated with rubber cement.

WOVEN CACHEPOT

As you can see, this woven container on the left literally lives up to its name, "hiding the pot," quietly and efficiently. It has been sprayed a recessive Krylon Khaki color, highlighted a bit with Illinois Bronze Accent Green Olive, but you could use any shade in those two lines that would complement the flowers you intended to put into it. Neutral colors like Satinwood and Wild Honey are always good foils for foliage, as is Mustard, but they are not always easy to find. One's hardware store, alas, is always at the mercy of the suppliers.

But, in any case, a woven pot is a big improvement over metal foil, being durable enough to last for years, and charming besides.

Coat the inside of the container around the base, where water might collect, with melted paraffin to protect it from chemicals.

Inasmuch as the Woven Cachepot is made on the same principle as the Bonbon Basket, I won't repeat the procedure here, but will only suggest that since it is taller, the ribs can be wider, say ¾". Keep the ribbon narrow, however, or it will buckle around the bend.

ROMAN COIN TUB

Once you get caught up in this canny thing, you will start scanning the shelves in the supermarket *just* for the containers, and you'll be in for some pleasant surprises. Many companies put a lot of thought into making them permanently appealing. This Roman Coin can on the right, for instance, put out by Maxwell House, was definitely intended to be more than a temporary container for coffee. Its royal blue and gold design is very striking, too striking perhaps for anybody's kitchen—better, perhaps, in the living room of a bachelor apartment. With lion's-head drawer pulls and *Juniperus wiltoni* cascading from it, it made a stunning display in our library for Earth Week, pointing up the essential usefulness and attractiveness of cans, and encouraging all of us to recycle them, or at the very least, rethink our habits of disposal.

If you would prefer a can equally elegant, but less obtrusive, see if your fish man doesn't receive scallops in a gold pail. Much like a paint pail, it has a refined, recessed lid and special sockets for the handles, which are ideal to fasten the lions' heads into.

Thinking of refinements, consider soldering feet on the pail, to give it a "leg up," as the British would say. That would literally lift it out of the tin can category, and it's easy to make Queen Anne feet. Try it sometime.

PINEAPPLE WALL POCKETS

Just between you and me, they are Marshall's Kippered Herring cans, but aren't they pretty? If you are handy with a brush you might want to paint them to look like real pineapples.

TOOLS

Snips
Long-nosed pliers

MATERIALS

Marshall Kippered Herring can
Paint remover and steel wool
Weldit Cement or Epoxy 220
Small gauge pull-chain or knitting needle
Sides of lacquered cans to match the interior of the kippered herring can

There's hardly any how-to to this, but I'll give you some pointers. First one, remove the lithographing *before* you open the can to enjoy its contents. It has been baked on and seems remarkably resistant to any paint remover, but it is sufficiently softened by it to make steel-wooling it easier. If you double the lid over completely, as in the lower right-hand pocket, you will only need to remove the lithographing on the *sides* of the can. If you prefer the two-tiered effect of the pocket on the upper left, you will have to remove some of the lithographing on the lower lid as well.

The only other trick is in opening the can. Make sure to start exactly in the middle of one side and go exactly to the middle of the other. Better mark it before you start and watch where the wheel of your can opener comes down.

Then there's one other thing. Because of the groove in it, lifting the lid up so that it bends *in the middle* is something of a problem.

Slip a knife in the opened end and very cautiously pry the lid up until you are able to work the tip of the long-nosed pliers in along the side to make the bend at the proper point. Be careful not to scratch the lacquer as you pinch the fold with the pliers.

After that, it's pretty clear sailing. If you plan to make a lot of these to sell at a church fair, you will want to trace the pineapple pattern onto Architect's Linen, rather than ordinary paper. (See the Pineapple Sconce for the pattern page 96)

Using a spiraled strip of tin as a border decoration is an old Mexican custom and seems appropriate here, but a small-gauge pull chain is attractive too. I used two widths for the spiral strips, one 3⁄16″ wide (around the outside of the pocket) and one 1⁄8″ wide on the inside. Wrap them around knitting needles or wire, whatever you have of consistent diameter.

Both the spiral and chain trim, as well as the pineapple backing, were glued on with Weldit Cement or Epoxy 220. Looking back on it, I think epoxy is better for this, especially if it is to be a sale item, where it will be picked up and examined before it reaches its destination.

JAPANESE LAN-SKETS

Either as lanterns or as baskets, you will find an unbelievable number of uses for these catchy containers based on the familiar Japanese lantern. Along the stone wall, around the pool, hung in the trees, or set on tables, you will find they brighten every imaginable situation. They can be large or small, plain or painted, lined or lighted. They hold candles or candy, dried flowers, fresh flowers, nuts or nasturtiums. Easy to make, here are the directions for the tiniest. Knowing the principle, you can carry on ad infinitum. (See color photograph, page 67)

TOOLS

Snips
Hammer and board
Awl
Letter-opener or knife
Long-nosed pliers

MATERIALS

Soup cans, preferably all-silver
Bottle caps, beer or soft drinks
Steel wool
Florist wire
Weldit Cement
Clear plastic pill vial *(optional)*
Dri-Mark pens *(optional)*
Krylon Crystal Clear Acrylic *(optional)*

One soup can will make *two* Tinies (for party favors, etcetera) or *one* Squatty (for citronella candle on the terrace).

Open up the can, hammer it flat, burnish it bright with steel wool.

Fold it in half lengthwise (see Scrolls and Candle Cups, page 17).

Mark off a line ⅜″ from the *open* edge and fringe the fold by making ⅛″ cuts down to the line (see diagram).

Hammer the fringe flat, working back and forth from the base of the cuts to the fold line.

If you want two little Tinies, cut the fringed strip in half, but if you need a Squatty, leave it whole.

Strike the awl holes as indicated in the diagram. Do this from both sides and hammer flat from the back.

Note: If this is to be a Christmas tree ornament in the form of a typical Japanese lantern with flowers, etcetera, paint the design on now with gaily colored Dri-Mark pens, *while the piece is still flat.* Let dry overnight and then give a protective coat of Krylon Crystal Clear Acrylic.

Open up the fold by running a letter-opener or knife *inside* the crease to separate the fringe.

Open it up like a book.

Turn it over and hold it lengthwise, crease up, from the top. Curve the lantern by *pressing it back over your thumbs* with your forefingers. Bring it around and wire it together top and bottom.

Note: For a Tiny, you will use the wire handle to fasten the lantern together; at the bottom you will need a cotter pin (see Paul Revere Lantern diagram, page 39). On the crease, bind the two ribs together with wire; pinch tight with the long-nosed pliers.

Perfect the circular openings top and bottom by pinching any unevenness with the long-nosed pliers.

Glue (or solder) the lantern onto a base of some sort. If a Tiny, use a flattened bottle cap upside down. If a Squatty, use the whole end of a soup can.

Braid a handle to go over the top (as in the May Basket).

TINY LAN-SKET

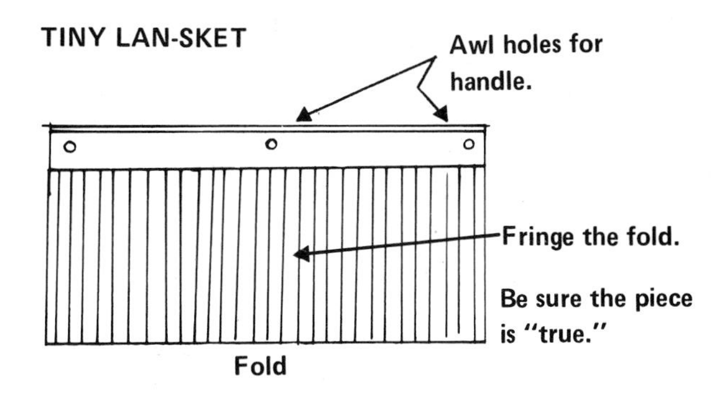

CONTINENTAL CACHEPOT

Cool and classical in feeling, this leafy cachepot frames both wild and cultivated flowers most effectively. Here you see Queen Anne's lace in company with Anthony Waterer spiraea, picking up the fuchsia border in the French plates. (See color photograph, page 101.)

Daisies are delightful in it too, being as snowy and crisp as the container itself. The pristine plastic placemats are in perfect harmony, even to repeating the classic oval form. (All that is lacking are Daisy Place-Card Holders, which you will find a little farther along in the book.)

As you can see from the procedural photograph, the cachepot is really masquerading as a container, being in actuality only a collar, cut to fit around an oval vegetable dish and set upon a wooden base. Made from a strip of tin, it is glued together with a 1½″ overlap. It would be easier, if you have to shop for the dish anyway, to find one that fits into the ham can perfectly, and thereby eliminate the overlap altogether.

TOOLS
Plain-edged snips
Hammer and board

MATERIALS
5-pound ham can
Shelf paper, scissors
Rubber cement
Krylon Flat White Spray Paint *(optional)*
Steel wool *(optional)*

Burnish the can with steel wool if you don't intend to paint it; otherwise, simply be sure it is clean.

Remove the bottom lid and both rims (see Removing Rims, page 14).

You will notice that there are two grooves in the side of the can. Cut the can down to the groove *nearest the edge* (see Cutting Down Cans, page 14). This will serve as a border for the base of your cachepot. Cut it accurately; it must be "true" in order to sit evenly on its base.

To make the leafy collar, cut a strip of shelf paper

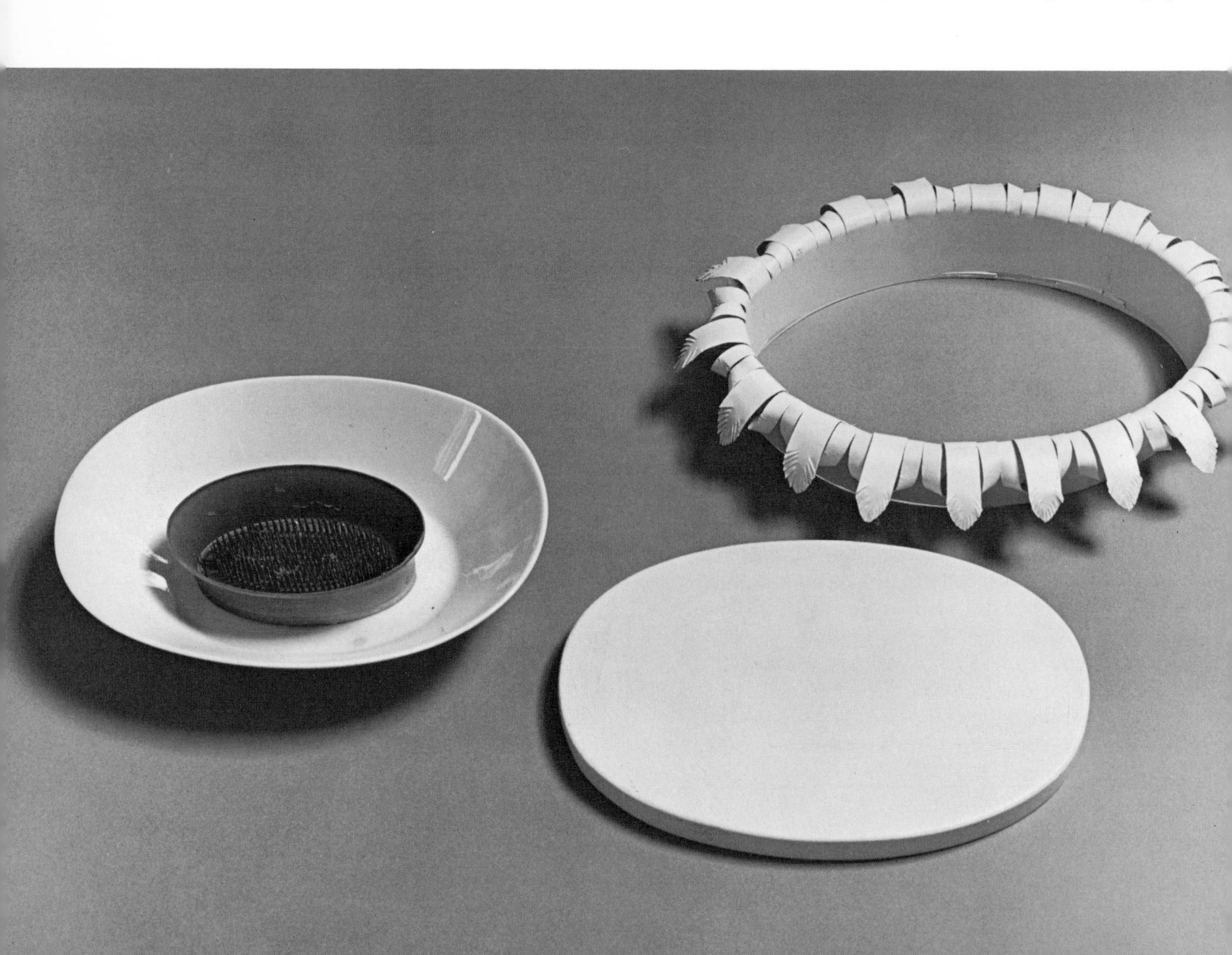

2¼" wide to fit around the top of the can (see Dividing Sides of Cans Into Equal Sections, Page 15).

Crease the paper in halves until the sections are about ¾" wide.

Coat the can with rubber cement and smooth on the creased paper. Cut along the creases; remove the paper and the glue.

Draw freehand Gothic arches on the end of *each* section and cut as indicated in the diagram, splitting every other section, and fringing the alternates.

Place the can across the corner of your hammering board and hammer each section flat without impairing the oval form of the container.

Curve each section toward you by pinching it between your thumb and forefinger.

Spiral the split sections to right and left as shown.

Spray-paint with Krylon Flat White Enamel, or whatever you wish.

The base is a pine oval cut with a sabre saw and sanded smooth. If you are a flower arranger, the chances are you already have one.

CONTINENTAL CACHEPOT

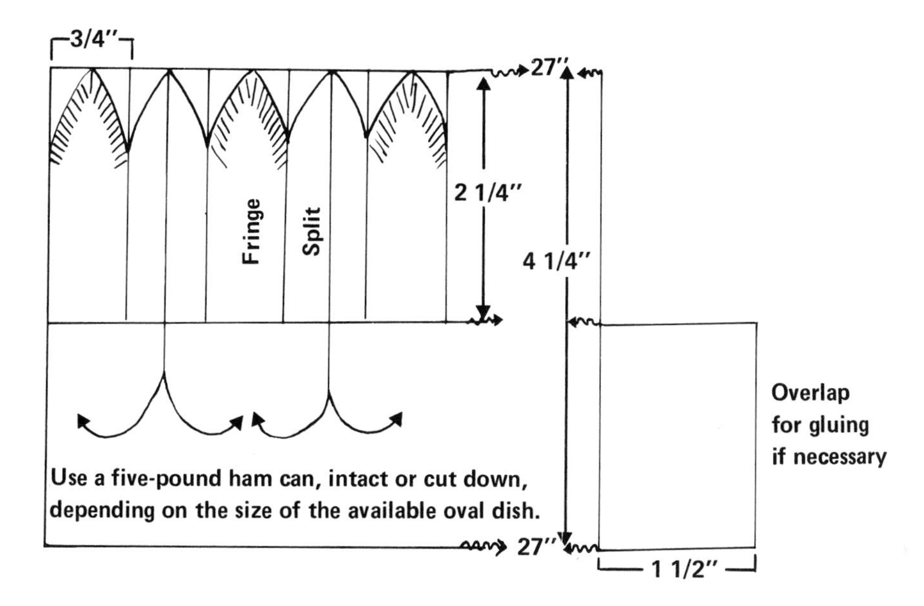

PIERCED POTPOURRI JAR

The intrepid American merchantmen who whisked around the Horn in the late 1700's, snatching seals and edible birds' nests from rocky archipelagoes to appease the Chinese, returned from the Orient with tea, spices, silks, and porcelains, introducing their countrymen to all sorts of exotic customs, including the burning of incense in exquisitely ornamented, latticed jars which you and I cannot approximate with the simple tools from our kitchen drawers. But we *can* produce a snug little container, with a pierced lid, ideal for incense or potpourri: A low one with large holes for incense, or a taller one with smaller perforations for potpourri. Except for the ornamentation, they take a matter of minutes, and no solder!

TOOLS
Snips
Ball-pein hammer and board
Awl
Compass
Chopping bowl

MATERIALS
2 *identical* cans
Talens Transparent Glass Paints
or Dri-Mark pens *(optional)*

Remove the top rim from both cans.

Using a compass to mark off the line (see Cutting Down Cans, page 14), cut down one can to within ½″ of the bottom rim.

Either fringe the remaining ½″ or divide it into coarser ¼″ sections, as shown in Fig. 2. Hammer the fringe flat from the *inside*. This will be the top of the jar.

To make the top fit over the bottom, bend the fringe out a little, place the lid in position, and squeeze the fringe tight with your hand. Tap it with the hammer until it lies even and flat. It's true, it will never fit as neatly as a commercial canister, but it *will* work, and it is a quick and easy way to make a small, covered container.

Pierce the top, making small awl holes for potpourri and larger ones for incense (you don't need many). As always, strike the holes first from the outside, then from the inside, and hammer flat from the inside.

Next, raise and round the pierced lid by beating it from the inside with the round end of the ball-pein hammer in your chopping bowl. Work diligently *around the rim*, where the metal is inclined to buckle, hammering until it is smooth and evenly curved. This will not only make the top look quite "proper," but will also ensure a truer seat when it is pressed onto the lower section.

Ornament the jar if you wish (Figs. 2-6). Paint a *bold* design on the lower section of the jar. Deep blue Dri-Mark and orange are particularly successful on all-gold cans (which one does come across occasionally). Talens yellow, green, and white are charming on *plain* tin. But, in any case, the colors should be chosen to harmonize with the setting in which they will be used.

PIERCED POTPOURRI JAR

DESIGN SUGGESTIONS

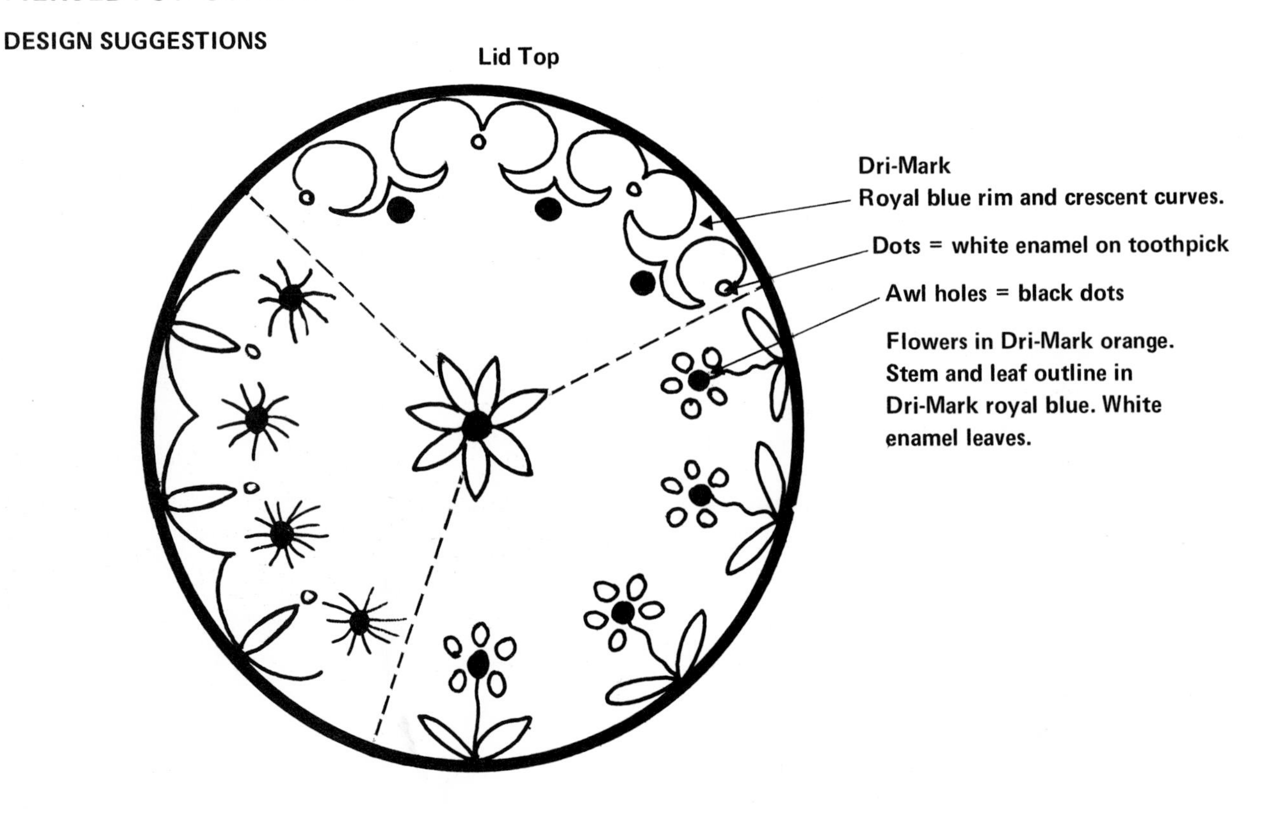

FIG. 1

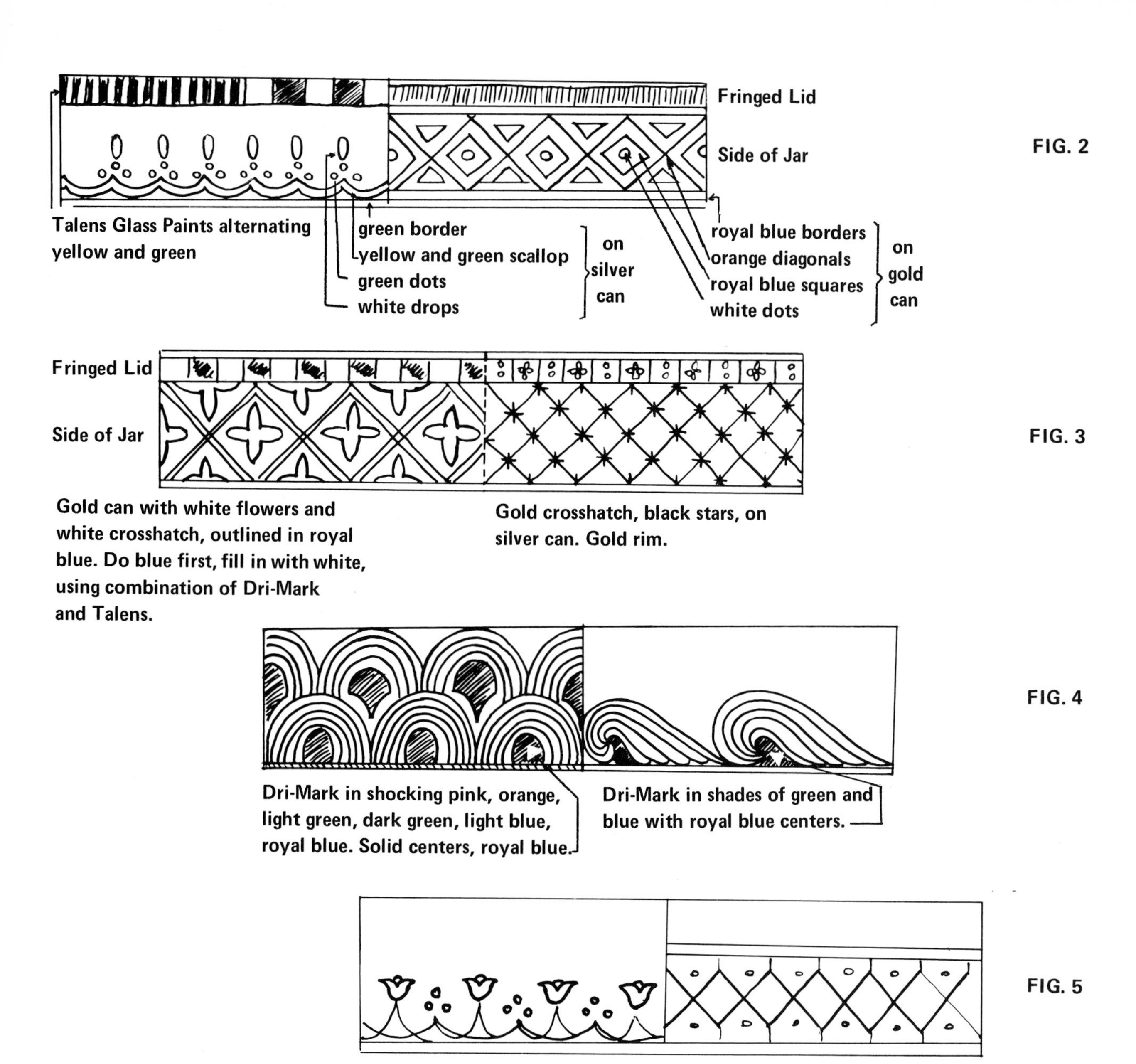

Note: Unless you are skilled with a brush, lines are more easily made with Dri-Mark felt-tip pens. White enamel, used sparingly, highlights a design most effectively.

These designs taken from Chinese export porcelain.

FIG. 6

SNAIL INCENSE BURNER

The original snail was designed by an artist in Taxco, Mexico's silver hilltown, for no purpose other than to delight the beholder. But it occurred to me that little ones would make beguiling place-card holders and incense burners. So here is my equivalent, with the horns converted into spirals to hold the joss sticks. It is fussy and requires solder, but once you get the hang of it, it will suddenly seem easy. Isn't that always the way?

TOOLS
Plain-edged snips
Ball-pein hammer and board
Awl and ruler
Long-nosed pliers
Round-nosed pliers
Metal hammering block
Pipe or broomstick
Propane torch
Asbestos mat
Vise

MATERIALS
1-gallon can, stripped of paint
Nu-Gold wire
Solder, Dunton's Tinner's Fluid
Felt-tip pen, turpentine

On the flattened side of a 1-gallon can, rule off a strip of tin 11″ long, grading it from ½″ at one end to 1¾″ at the other (Fig. 1.). Cut and hammer flat from both sides.

SHELL DIAGRAM

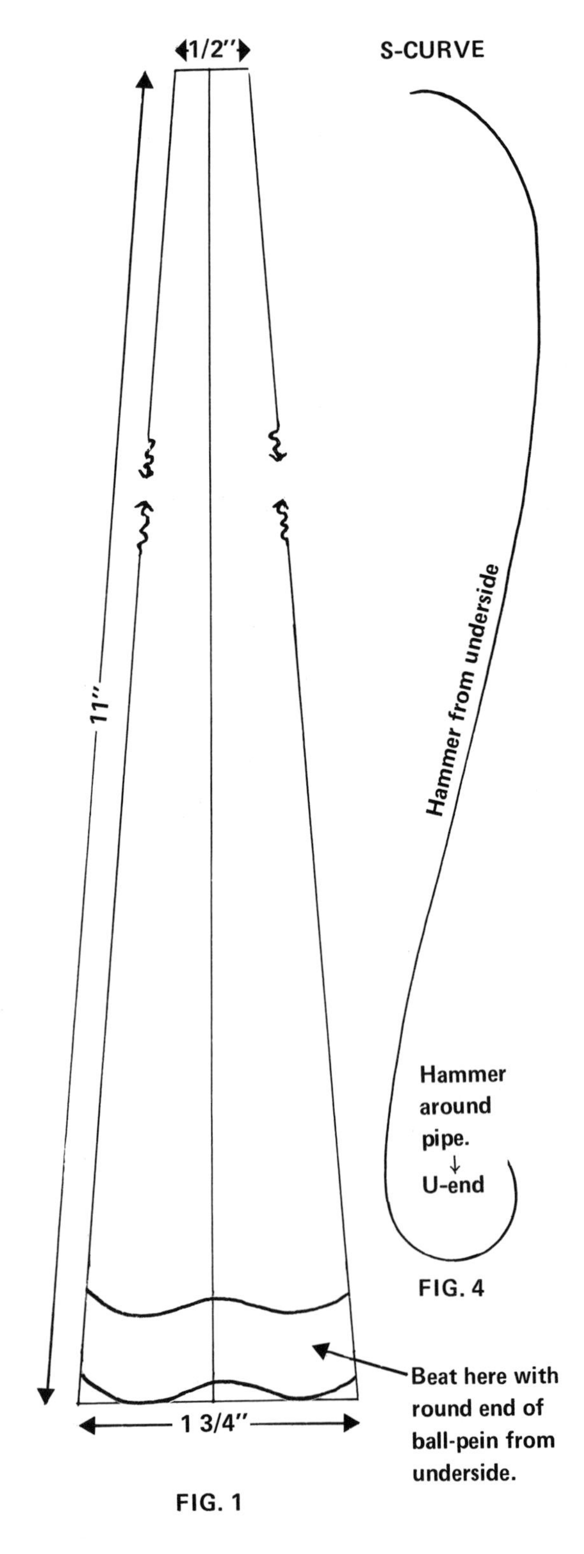

FIG. 4

FIG. 1

Cut another, smaller piece, 3½" long, grading it from ¼" at one end to ¾" at the other (Fig. 2). Hammer flat also.

With a felt-tip pen, mark the curves on the *broad* end of *both* pieces as indicated. Cut, and hammer flat.

Strike the two small awl holes in the Snail's foot (Fig. 3). Strike from both sides and hammer flat from the back, as always.

SHAPE THE FOOT

The foot should more or less resemble a half-cone, which will be used *curved side up, open side down,* with the shell riding on top of it. (Can you make that out in the photograph?)

To get it to curve on the sides, beat the foot on a *flat* board with the *round* end of the ball-pein hammer. If it tends to curl up *end to end,* turn it over and tap it down gently (Fig. 3).

When you've done all you can to curve it with the hammer, grasp the narrow end of the foot with the round-nosed pliers, and bend the metal around to form a peak. To accomplish this you may have to squeeze the metal *around* the round-nosed pliers with the long-nosed pliers.

Then work along the sides of the foot with the round-nosed pliers, encouraging the sides to curve up. Try to avoid rippling the edges; your gastropod should have a firm footing.

FOOT DIAGRAM

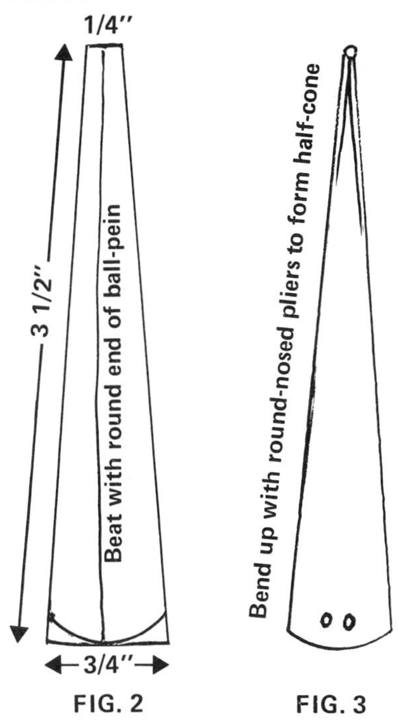

FIG. 2 FIG. 3

SHAPE THE SHELL

Before it is coiled, the shell strip must first be made to resemble an elongated S-hook (Fig. 4). Form a U on the broad end by hammering it around a pipe or broomstick (see Scrolls and Candle Cups, page 17).

Turn the strip over, and hammer the full length of the strip, from the narrow end *down to* the U, until the metal comes up in a really pliant, reverse curve. You cannot hammer it too much; the more it curves up under the hammer, the easier it will be to coil.

Start the coil with the round-nosed pliers by twisting them around the *narrow* end of the strip *tightly.* Switch to the long-nosed pliers and squeeze that initial coil into as *tiny a tube* as you possibly can. This is the vortex, the very heart of the spiral, and it should be as tight and perfect a whorl as you can make it.

The rest of the spiraling shell is formed, almost entirely, by pinch-pinch-pinching toward you with the round-nosed pliers, holding the U-end in your helping hand. (If you've ever made the Penobscot Indian Bells in *Tincraft for Christmas,* you know what I mean.) It involves a slight backward twist of the wrist along with the pinching. The forward movement is almost imperceptible, something like 1/32" for each pinch. *Very deliberately* you pinch forward, and *very gradually* your snail spirals off behind. When you once catch on to it, you won't even have to think about it; your hand and wrist will do the job for you.

Don't worry about the lopsidedness of your snail's convolutions; you can poke them back in if they stick out too far, but they *should* be a *little* lopsided or your snail won't seem real; he'll just be a loose roll of tin.

When viewed from the side, there should be altogether five convolutions, about ¼" apart. If you've inadvertently made any creases as you pinched, flatten them with the long-nosed pliers. And if that process enlarges the spiral too much, squeeze it shut with your hand and pull it out again with your fingers. Play with it until it spirals pleasingly.

Off-angle the center whorl by twisting it up and forward. Amazingly, this will endow your snail with vitality and movement.

Finish off the shell by shaping the U-end. This is the orifice from which the snail's head (foot!) emerges, so it must curve up in the middle and down and around on each side. Therefore, turn the shell upside down and beat around the curves with the round end of the ball-pein hammer. See how it frames the foot (head!)?

MAKE THE FEELERS AND HORNS

This is the ingenious part, and I'm very pleased with it. The feelers are purely decorative and contribute half the charm to our creature, but the "horns" are impressively functional and can be readily adapted to fit *any* size of joss stick! That is rather a neat trick, considering the different sizes they come in, but let me take you step by step.

For the feelers, cut a 6″ length of Nu-Gold wire. Grasp it in the middle with the round-nosed pliers and bend the ends together (Fig. 5). Send the ends up through the awl holes in the foot.

Grasp the feelers with the round-nosed pliers right next to the foot where they emerge, and with your fingers, *bend them forward over the pliers.*

If they have any kinks in them, straighten them out with the long-nosed pliers.

Coil each feeler with the round-nosed pliers (see Baubles and Bangles for similar design, page 163) and hammer them *flat* on a metal hammering block.

Solder them in place from inside the foot. (Note: you may want to leave this loose for the time being until you've made the horns, and then do all the soldering at once.)

For the spiral horns, cut a narrow strip of tin, something between 1⁄16″ and ⅛″ and 4″ long. Using a length of Nu-Gold wire as a form for the spiral, wind each end of the tin strip around the wire, *leaving a ¼″ unwound section between them* (Fig. 6). Solder the horns in place on top of the foot behind the feelers, having first tinned both pieces where they touch. The sole reason for soldering this piece is because of the stress placed on the horns *when you wind them up or down to accommodate the different-sized joss sticks.* If a stick isn't held snugly in place, it won't stand up straight, but will, instead, flop over sidewise and shed ashes all over the table. With this clever device, our snail carries his incense proudly and only anoints himself with ashes.

One last thing to do. Solder the snail's shell onto the foot. Here again he comes alive, if his shell is placed *at an angle* on the foot, rather than being centered on it. To be sure, the head-end *is* centered, but the tail-end is positioned *well off to the side.*

Note the two points where the shell and foot touch. Tin those points on both the shell and the foot. Clamp them into position in the vise. Apply heat and perhaps the tiniest snick more solder, and you've got it made. Congratulations!

FEELER DIAGRAM

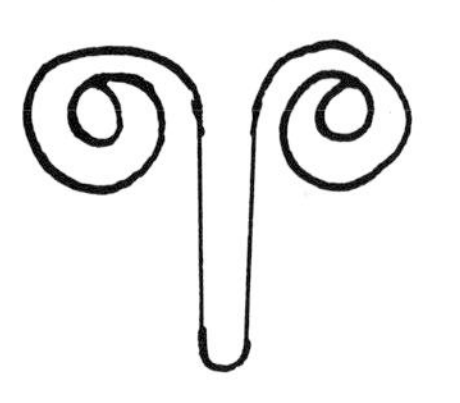

FIG. 5

HORN DIAGRAM

FIG. 6

DAMASK ROSE INCENSE BURNER

As practical as it is pretty, this damask rose can hold either cones or joss sticks, and without the leaves and handle, takes only minutes to make and no solder!

TOOLS
Snips
Ball-pein hammer and board
Awl
Chisel *(optional)*
Propane torch *(optional)*
Asbestos mat *(optional)*

MATERIALS
5 matching lids, grading from large to small, with approximately ¼″ difference in diameter
Bottle cap
Florist wire
Solder, Dunton's Tinner's Fluid
Narrow strip of tin for spiral joss stick holder *(optional)*
Tracing materials: paper, scissors, pencil, rubber cement *(optional)*
Side of matching can for leaves *(optional)*
16-gauge wire or brass welding rod for stem and handle *(optional)*
Sandpaper *(optional)*
Paints *(optional)*

If you are going to make *just the rose,* use *larger* lids than if making the whole spray, grading them from 4″ to 1½″ in diameter, with approximately ½″ difference between them. *With* the leaf spray, range the lids from 3¼″ to 1½″, with ¼″ difference in diameter between them.

Whatever the size, the lids should be *all the same color.*

THE ROSE

Hammer the lids flat.

Strike *two* awl holes, side by side, in the *center* of each lid (from front, then back, and hammer flat from the back).

Divide each lid into *eight* equal sections (see Dividing Lids Into Equal Sections, page 15), except the smallest one, which should be divided into *four* sections.

Round each section as shown in the diagram. (See also Apothecary Rose technique, page 48, for fast rounding.)

Beat each individual petal in a bottle cap with the round end of the ball-pein hammer. Beat the larger petal layers *from the back,* the smaller *from the front.*

Cut a 4″ strip of tin, ⅛″ wide, for the joss stick holder. Wind *each* end around a very fine knitting needle or wire, leaving a ½″ *unwound* section in the middle.

Assemble the rose by sending a length of florist wire up through one set of holes in the petal layers, catching the joss stick across the middle before sending it back down through the other set of holes. Twist the wire tight. If you would feel more secure, you could put Weldit Cement between the layers; otherwise, that's all there is to it! Unless you want to add the leaves.

THE LEAVES

If you are only going to make one burner, it is not necessary to use Architect's Linen for the patterns. Simply trace the two leaf patterns onto any kind of paper, and don't even bother to cut them out.

Apply rubber cement to the side of the can; smooth the paper on; and cut right through the paper. Be careful not to tear the pattern as you remove it, because you will need to use it once more.

DAMASK ROSE INCENSE BURNER

LEAF PATTERN AND DIAGRAM

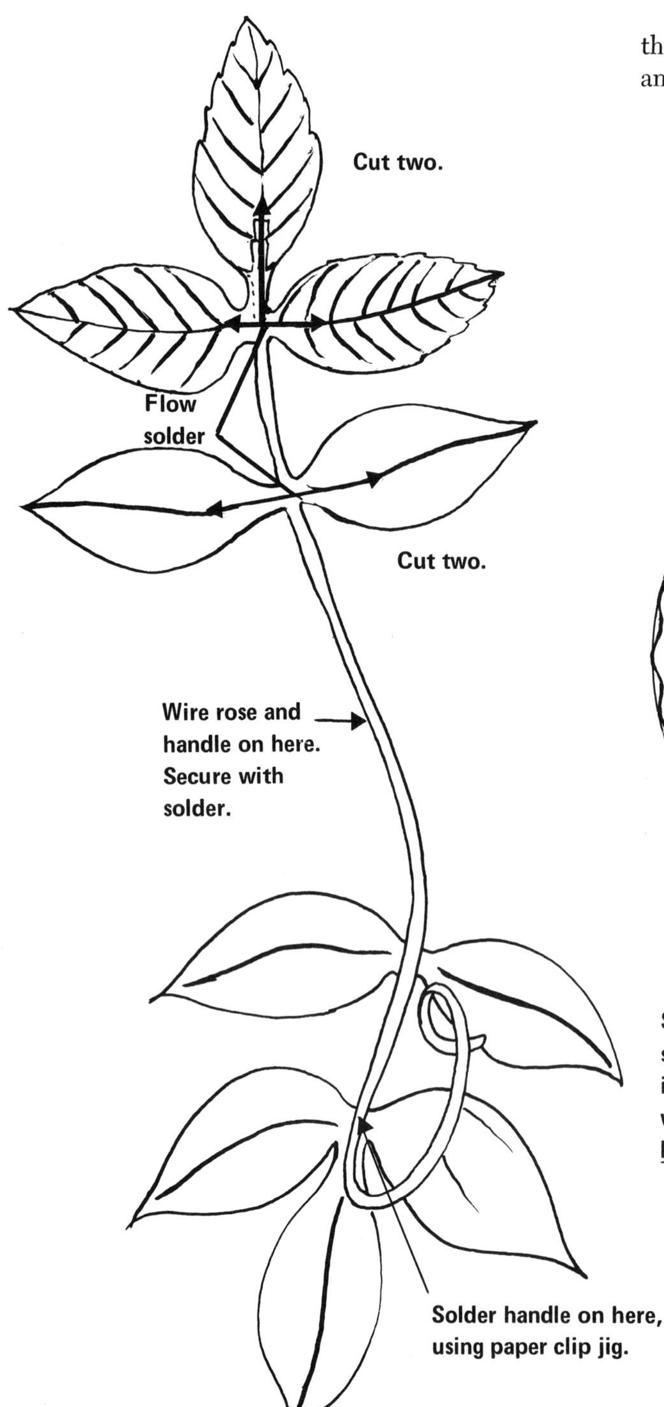

Vein the leaves (unless you prefer them plain), using the awl to gain a natural effect, or a chisel for a stylized effect.

Clean the areas to be soldered with sandpaper.

Cut a 6″ length of 16-gauge wire (if silver leaves) or brass welding rod (if lacquered leaves) and put a slight S-curve in it as shown in the diagram.

Assemble the leaves and wire on the asbestos mat and solder.

Add a handle if you wish; bind the rose around the handle and leaf stem; secure it as indicated with an additional drop of solder.

PETAL LAYER DIAGRAM

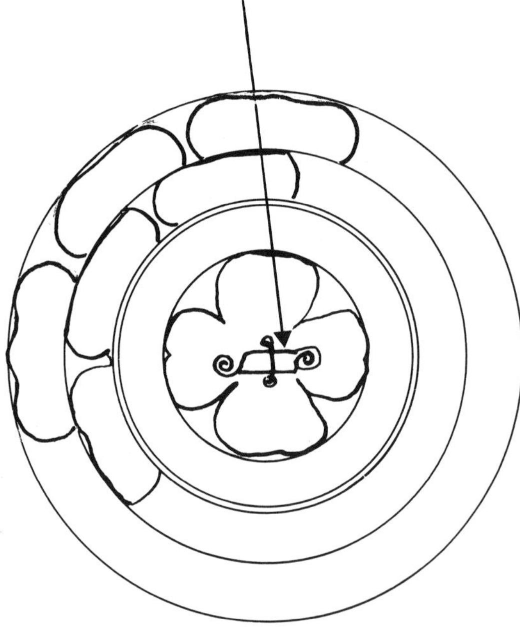

Strike two awl holes in center of each lid. Divide all but smallest of lids into eight equal sections, the smallest into four. Round sections and beat in beer bottle cap with round end of ball-pein hammer: large layers from back, small ones from front.

PART FOUR
all around the house

LIGHTING THE WAY

Now that candles are a billion-dollar business, it would seem that most people agree with my mother, who brought us up to have at least one in every room in the house. This was not because we spent summers on the farm where there was no other source of night light, but because she felt there was nothing more conducive to human communication than the magic created by candlelight. Conditioned in this way, there is nothing remarkable about the candle count in my house, numbering twenty-six in the living room (partly for chamber music), nineteen in the dining room, and four or five in each of the other rooms in the house before ever moving out into the garden. And speaking of the garden, notice how happily a sconce can hold flowers.

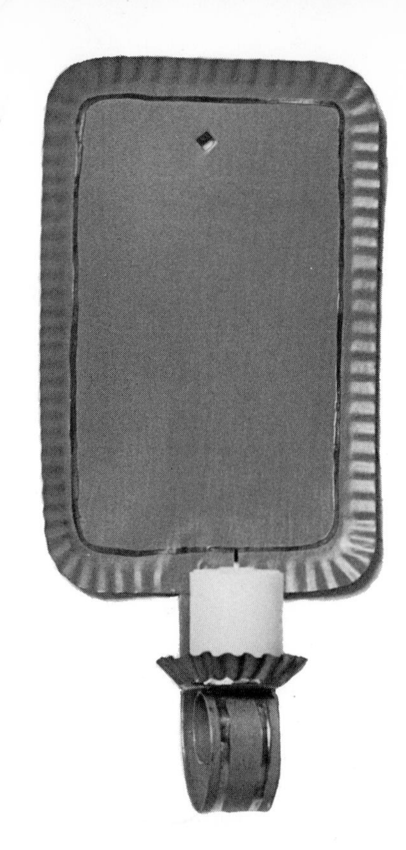

TWO EASY, EASY SCONCES

A crank of the can opener and a spot of glue, and you've just about got it made.

TOOLS

Snips
Ball-pein hammer and board
Awl
Vise
Short length of pipe or broomstick

MATERIALS

Whole end of 3-pound coffee can
Bottom lid of 1-gallon can
Strips of tin for scroll and candle socket
Lid from small frozen juice can for crimped cup, or
  whole end of soup can for drip cup
Weldit Cement
Florist wire
Pull chain *(optional)*
Paints and brush *(optional)*

Except for the scroll, there is literally nothing to this: it's practically all been done for you by the manufacturer. In other words, a lid stamped with concentric circles is an essentially pleasing form, and by raising and texturing the center section, and gluing on a shiny bright length of chain, you have the makings of a respectable sconce.

Use the whole end of the 3-pound coffee can, including the rim (see Removing Rims, page 14). This not only frames the circle, but provides extra gluing surface for the chain.

Beat the center section with the round end of the ball-pein hammer from the *back* on a *flat* board, being careful not to spoil the perfection of the concentric circle by hammering over the edge.

Cut the pull chain to fit; run Weldit Cement around the edge and allow to become tacky. Press pull chain in position.

Glue scroll (see Scrolls and Candle Cups, page 17) and wired tab on the back for hanging (see diagram on page 95, Fig. 2), and you've finished.

A simple, typically Early American device for putting finish and flourish on the edge of a sconce is crimping. This is easily accomplished with the round- or long-nosed pliers and a twist of the wrist to right and left, working clockwise around the edge. (See Scrolls and Candle Cups.)

An even simpler device is a border of chisel marks, and the very simplest sconce of all (which you will see if you go to Old Sturbridge Village in Massachusetts) is an unadorned tin oval. This last form is so pure, it achieves surprising elegance.

The second Easy Sconce you see here has been painted Old Yellow (mustard) with a gold border line.

PLUME-POPPY SCONCE

Ten feet tall and easily as wide, my perennial plume-poppy plant dominates the tool shed. In true poppy fashion, the leaves are silver-gray on the underside, and as large as dinner plates. The foot-long, pinky-beige spikes apparently possess a heady elixir all their own, because the bees, intoxicated, buzz wildly about them. The *Bocconia cordata* was my grandmother's favorite, and it is mine too. And small wonder! Any plant so spectacular and so controllable *should* command our admiration. Here is one for you to enjoy indoors.

TOOLS

Plain-edged snips
Hammer and board
Awl
Screen installation tool

MATERIALS

1-gallon can
Tracing paper or Architect's Linen, scissors, pencil
Rubber cement
Carbon paper and ballpoint pen
0000 steel wool (very fine)
Whole end of soup can
V-8 Juice can
Weldit Cement
Florist wire
Krylon Crystal Clear Acrylic Spray *(optional)*

PLUME-POPPY SCONCE

PATTERN

Trace off the pattern, including the veins (onto Architect's Linen if making more than one sconce); cut out and apply with rubber cement to the polished side of a 1-gallon can.

With plain-edged snips, cut painstakingly around the leaf, swooping around the deep curves, and taking tiny little bites around the shallow ones (see Cutting Out Patterned Designs, page 16).

Remove the pattern and the glue. Hammer around the edge of the leaf and trim off any unevenness until it is rippling smooth. Hammer again.

Vein the leaf, preferably with the screen installation tool, which makes it so easy you can do it by eye.

Otherwise, you will have to vein it with an awl. In which case, reposition the pattern on the leaf; slip a piece of carbon paper under it; and trace on the veins with a ballpoint pen.

Score the veins lightly the first time, holding the awl as if it were a pencil. Then go over the lines again, this time working *backward* from the outer edges of the leaf, so you can see exactly where you are going. Hold the awl like a weed-digger, and press as hard and firmly as possible, guiding the tip of the awl with the fingers of your helping hand. Pay special attention to those points where one vein meets another; they should flow together. The Mexicans do all such scoring on several thicknesses of inner tubing, but a soft pine board is just as good.

When all the veining is done, hammer the leaf all over, between the veins, with the flat end of the hammer from the back. This will give your leaf a soft and supple appearance. Enhance it even more by polishing it with *0000* steel wool.

Make the scroll of rich gold from a V-8 Juice can to contrast with the satiny-silver leaf (see Scrolls and Candle Cups, page 17), and glue it with Weldit Cement or Epoxy 200 to the back of the leaf. Also glue on a wired tab for hanging (see diagram on page 95, Fig. 2). A 4″ candle seems right for this sconce.

OAK LEAF SCONCE

Anyone passionate in pursuit of knowledge of our American heritage should make a pilgrimage to Winterthur. I say "pilgrimage" because it is certainly with awe and a kind of reverence that one views this choicest and most extensive collection of Americana in the country. Once the residence of Henry Francis du Pont, it is now a museum consisting of almost two hundred period installations, representing American decorative arts from the seventeenth century through the early nineteenth century.

Presumably, you as a serious student would want to tour the Main Museum, which can be visited by only a limited number of people each day and therefore requires your writing the reservations office (Winterthur Museum, Winterthur, Delaware 19735) well in advance. You should also allow time to roam through the gardens there when they are open, and at Longwood, another incomparable du Pont residence a few miles away. As Americans, we are enormously indebted to this family of French origin who have contributed so much to their adopted country's culture.

The Oak Leaf Sconce lights the Pottery Room at Winterthur.

OAK LEAF SCONCE
PATTERN

TOOLS

Plain-edged snips
Awl
Hammer and board
Screen installation tool
Long-nosed pliers
Screwdriver
Short length of pipe or broomstick

MATERIALS

2-gallon oblong can
Architect's Linen, pencil, scissors
Rubber cement
Ruler
Posey Clay
Felt-tip pen, turpentine
Ballpoint pen, carbon paper
Krylon Flat Black Spray Paint

A 2-gallon oblong can accommodates this 23″ sconce generously, but make a point of positioning the leaf pattern so that the base of it coincides with a bend made by a corner in the can. For though you will start the sconce by opening up the can and hammering it flat, still, a trace of a former curve in the can will be visible, and it would be best used where a curve in the sconce is called for. You may already have noticed that the oak leaf tilts forward, in the manner of many early reflector sconce backs. (A sufficiently large round can could also be used for this sconce.)

You only need trace patterns for the leaf and star. The rest of the sconce can be ruled off with a felt-tip pen.

CUT OUT THE SCONCE

Trace the leaf design, including the veins, onto the dull side of a piece of Architect's Linen, and apply with rubber cement to the unlithographed side of the can. (It is not necessary to remove the lithographing, if any, because the sconce will be spray-painted flat black in the end.)

With ruler and felt-tip pen, mark off the shaft to an overall length of 23″. Cut out with plain-edged snips.

Round the base of the shaft as indicated in Fig. 1.

Hammer the whole sconce around the edges from the back side. Make any improvements necessary in the outline of the leaf, and hammer flat again.

VEIN THE LEAF AND BORDER THE SHAFT

Reposition the leaf pattern, slip a piece of carbon paper underneath it, and trace on the veins with a ballpoint pen.

Score along these lines with the awl, or the convex wheel of the screen installation tool (see Veining, page 16).

On the shaft of the sconce, score a border line ⅛″ from the edge, preferably with the screen installation tool, using the convex wheel from the front and the concave wheel from the back. Two little dabs of Posey Clay on the bottom of the ruler will help to keep it in place on the shaft and make the border line true. Press hard on both the ruler and the tool to make a clear impression and provide the shaft with a pleasing frame. Like any rib, this border line also helps to make the sconce back rigid.

THE STAR

Apply the star pattern to the *back* of the sconce and strike all the awl holes through the linen, *except* for the large center hole, which should be struck from the front. Drive the awl down through the hole right up to the hilt. Do the same from the back and hammer flat. The sconce will hang from this hole (unless you prefer to wire it from the back).

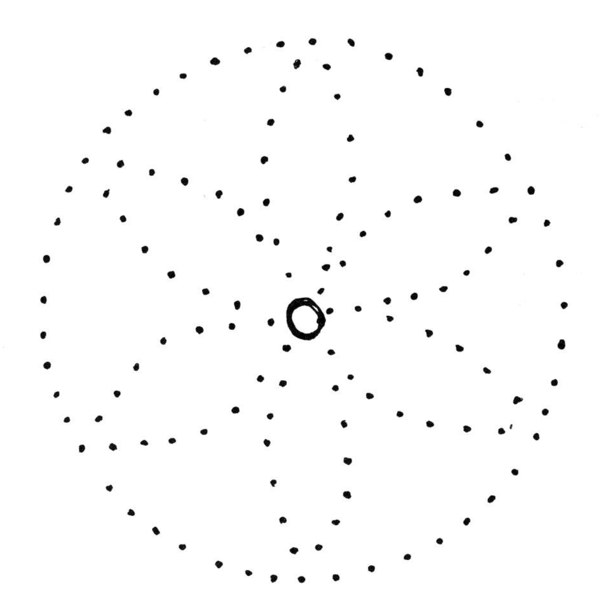

THE PAN RIM

It is absurd to ascribe human emotions to inanimate metal, but I am almost driven to think a pan rim is intentionally perverse. No matter how carefully I measure it, it turns out to be too short or too long.

SHAFT

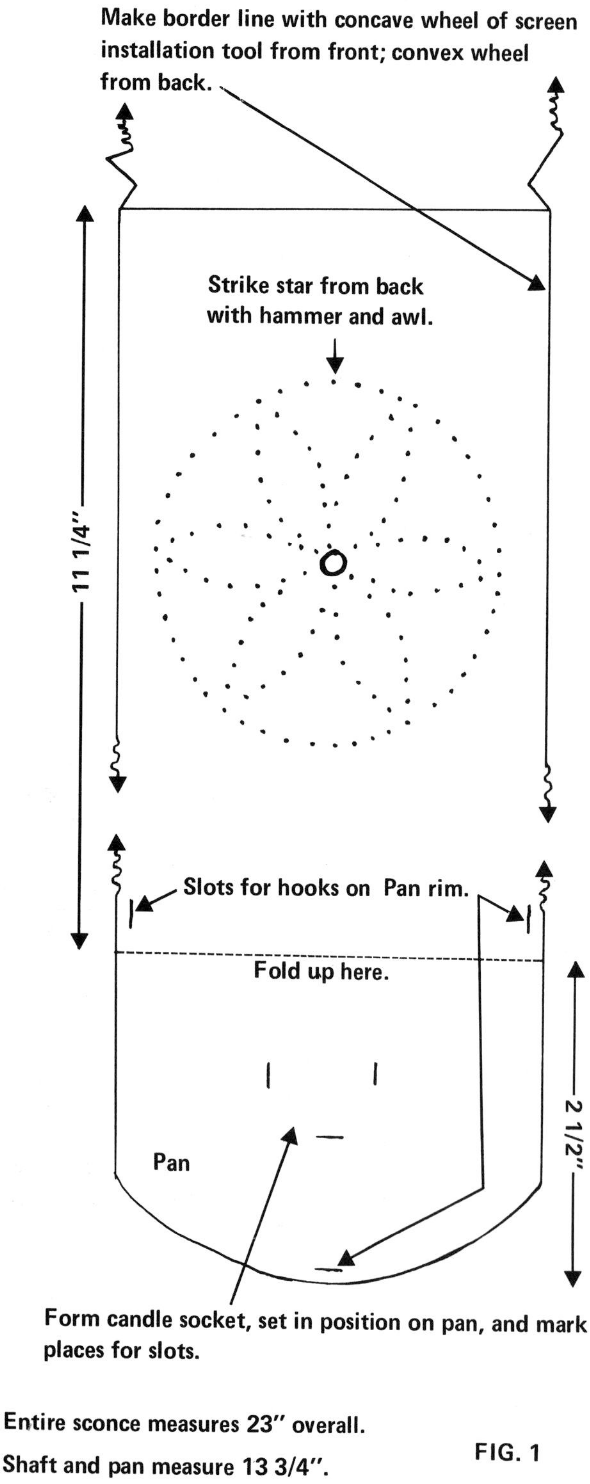

Make border line with concave wheel of screen installation tool from front; convex wheel from back.

Strike star from back with hammer and awl.

11 1/4"

Slots for hooks on Pan rim.

Fold up here.

2 1/2"

Pan

Form candle socket, set in position on pan, and mark places for slots.

Entire sconce measures 23" overall.
Shaft and pan measure 13 3/4".

FIG. 1

So, be prepared to hurl invectives, but don't let it depress you. And cut it long, rather than short, because you can always trim it down where it wraps around the back of the sconce.

Cut a strip of tin for the pan rim 8½" x 1½". Make tabs by cutting away ¼" on three sides, as indicated by the dotted lines in the diagram (Fig. 3). Better line it off carefully with felt-tip pen and ruler. The tabs should be a solid ¼" deep, no less. Be sure to center the one on the long side; the others won't show and don't matter quite so much.

Score the border lines with the screen installation tool, the first ⅛" from the top of the pan rim and the second ⅞" below it.

Turn the piece over and score another line with the awl exactly ⅛" from the lower edge (Fig. 3). Be exact about this. Cut little wedges as shown in the diagram just *to* that score line. Perfection counts. You don't want any little nicks to show at the base of the pan rim after you bend that flange under. Cut the notches labeled *y* and *z*, at either end.

With the flat of the hammer, beat the pan rim on the *inside of the wedged section,* so that it curves up smoothly. Leave both ends flat.

With the long-nosed pliers, bend the flange down at right angles. The scored line on the inside should help you to get the bend perfectly true, but it's not easy. If you inadvertently bend it too far, simply hammer it flat and start over.

In order to wrap the pan rim around the back of the sconce, bend both ends of the rim back at right angles with the long-nosed pliers.

Once you've successfully shaped the pan rim, bend the base of the shaft up (Fig. 1) and wrap the pan rim around the curve to determine where the slots for the tabs should be placed. Mark them carefully and strike them in with hammer and screwdriver. In order to fit the tabs into the slots, bend them with the pliers.

THE CANDLE SOCKET

Cut out a strip 1¼" x 3¼" and trim along the dotted line to form tabs (Fig. 2).

Hammer the strip around a short length of pipe or broomstick (see Scrolls and Candle Cups, page 17) until it overlaps ⅛". Glue the overlap and secure with paper clips until dry.

Set the candle socket in the pan and mark the places where the tabs touch. Strike the slot with hammer and screwdriver. Insert the candle socket and bend the tabs against the underside. Hallelujah!

Spray-paint the sconce flat black—with a vengeance!

CANDLE SOCKET

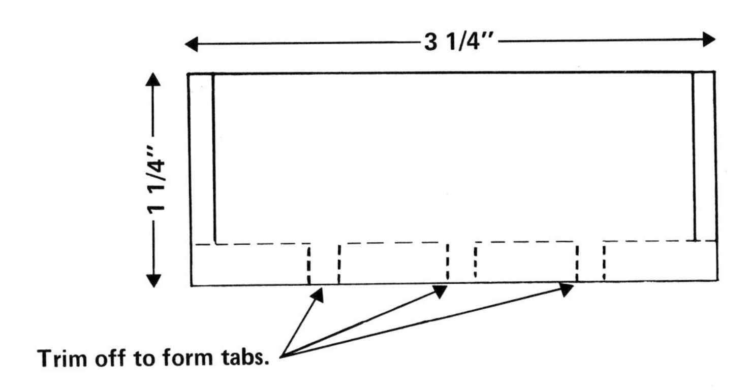

FIG. 2

PAN RIM

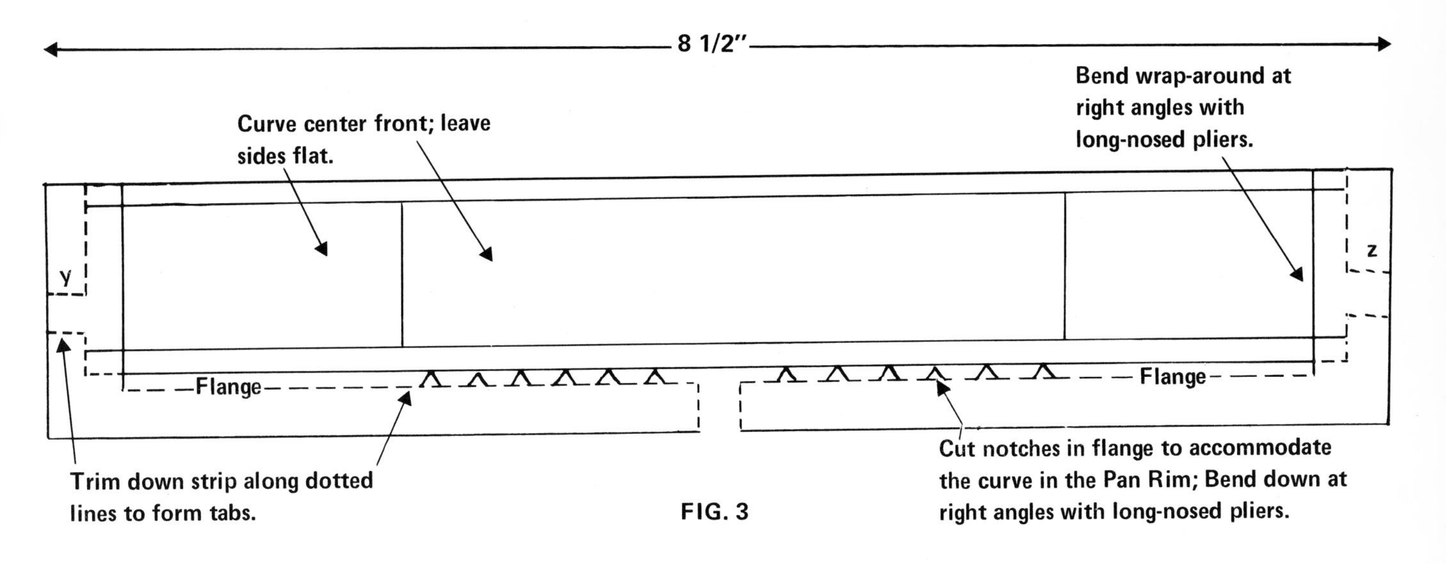

FIG. 3

FRENCH PROVINCIAL BOWKNOT

Aside from its simple sophistication, this is a very useful bowknot. It first came to my attention when a friend brought me the Damask Rose Sconce for repair. Because it takes nothing unto itself, and serves its purpose with easy grace, it frequently fits into a composition when a more elaborate bow would not. It's easy to make, too, if you are fastidious in cutting out the pattern.

TOOLS
Plain-edged snips
Round-nosed pliers
Hammer
Broomstick or pipe
Propane torch *(optional)*
Asbestos mat *(optional)*

MATERIALS
Unribbed side of can
Architect's Linen, scissors, and pencil
Rubber cement
Solder, Dunton's Tinner's Fluid *(optional)*
Weldit Cement *(optional)*

In the loop pattern you will notice that the ends are higher than the section on the fold. That is because of the flare in the loop. The higher the end, the greater the flare; something to remember should you want more flare at one time than another.

Trace the loop and bow-end patterns on a fold of Architect's Linen and glue the folds of linen together with rubber cement to ensure accuracy and perfect symmetry when cutting out.

Coat the unribbed, burnished side of a can with rubber cement and lay the opened patterns on carefully so there is no distortion in the design. Cut out with plain-edged snips. Hammer flat.

Find the center of the loop and mark it. Bend the ends of the loop around a broomstick and bring them around to meet at the center mark.

I like to ripple the bow a bit in front on either side of the center, and usually use the round-nosed pliers, waggling them to right and left, but you can bend the metal back and forth over the broomstick if you prefer. You'll get a smoother, but somehow less realistic, ribbon.

Ripple the bow-end similarly.

If you have decided to glue the bowknot together, rather than solder, apply Weldit Cement to all the layers. Allow it to become tacky. Slip the bow-end inside the loop and secure with a paper clip till dry.

Cut the knot strip. Hammer flat. With the round-nosed pliers, bend the ends up along the dotted lines. Apply Weldit Cement to the center inside and press it onto the *front* of the bow.

You only need a tab if you are binding a collection of stems together, as in the Damask Rose Sconce. In which case, lay it across the back of the bow and bend the knot strip over it. Apply glue, or solder, to the overlap.

Do you recognize the Oak Leaf from Winterthur?

LOOP PATTERN

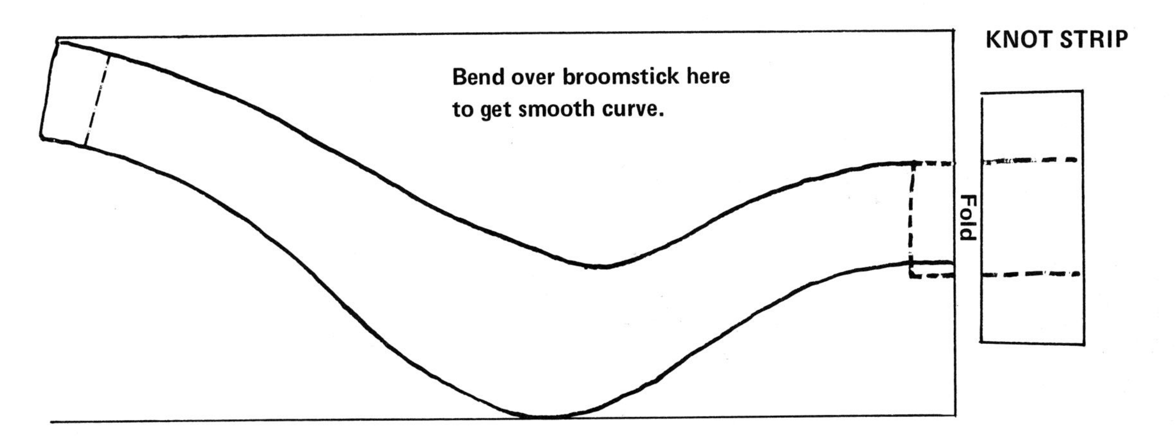

ASSEMBLY

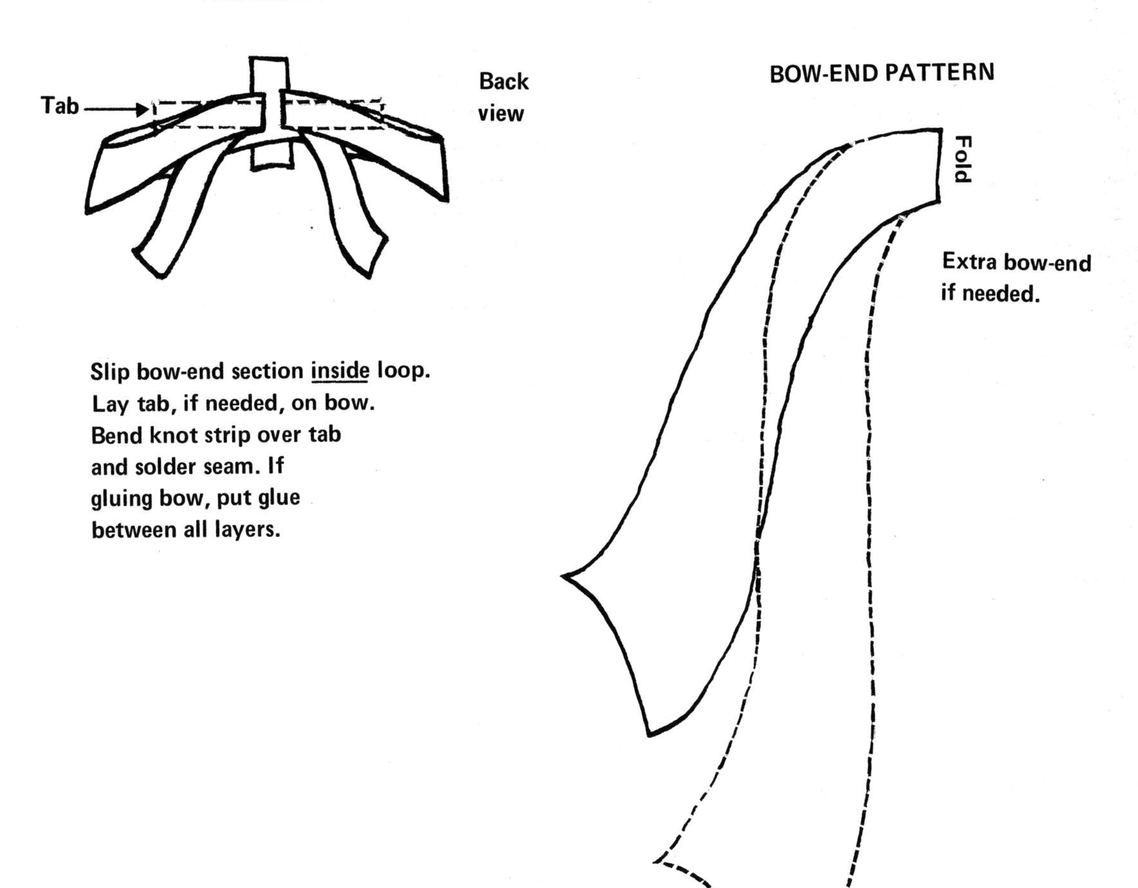

Slip bow-end section inside loop.
Lay tab, if needed, on bow.
Bend knot strip over tab
and solder seam. If
gluing bow, put glue
between all layers.

PENNSYLVANIA DUTCH TULIP SCONCE

No doubt about it, those Pennsylvania Germans had an eye for designs to gladden the heart: their birds are so bright and their tulips so sprightly. As you can see in the color photograph on page 101, my Spider-plant babies are sitting snugly under one of their tulips and rooting away merrily.

TOOLS
Plain-edged snips
Hammer
Awl

MATERIALS
1-gallon can
Whole end of soup can
V-8 Juice can
Tracing paper or Architect's Linen, scissors, pencil
Rubber cement
Weldit Cement
20-24 gauge florist wire
Paints or Dri-Mark pens
Mirror *(optional)*

TRACE THE TULIP

In this design, where symmetry is everything, ensure the accuracy of your pattern by *pasting the fold of tracing paper or Architect's Linen* (if making more than one) *together with rubber cement before cutting it out,* so that it won't slip but will make a perfect "mirror image."

Be sure, when positioning the opened pattern on the tin, that the petals are perfectly erect and uniformly distant from each other.

Take great pains in cutting the sconce out. Shave off any slight exaggerations or deviations from the pattern. It's amazing how the eye will detect discrepancies, even from a distance.

Do not attempt to cut the holes in the handles on the basket. Paint areas to simulate holes in the handles.

THE SCROLL AND DRIP CUP

If your color scheme calls for a rich gold scroll, make one from a V-8 Juice can (see Scrolls and Candle Cups, page 17. Otherwise, any sturdy can will do, especially if you intend to paint.

As for the drip cup, use the whole end of a soup can and hammer it from the inside on a flat board with the round end of the ball-pein hammer. This will make a nice shallow saucer to hold either a votive candle or a container of flowers.

Flowers and water, however, are heavy for anything as flexible as a scroll. So, before gluing the cup on the scroll, strike *two tiny awl holes under the rim* at the back of the cup (Fig. 1) and *two more* through the corresponding point on the scroll directly behind it. Then later, when the cup has been glued to the scroll and the scroll to the back of the tulip, you can wire the cup to the scroll and the flowers will be in no danger of spilling forward.

Incidentally, the large cap on a can of spray paint makes an excellent container for flowers, and fits nicely into the drip cup. Some caps even come with a circular section in the center where you can place a candle and then surround it with flowers.

Make the hook for hanging, wiring it onto an extra-long strip of tin to reinforce the back of the sconce.

PAINT THE SCONCE

You will find two color schemes suggested in the diagram, but of course, your own decor will decide the matter. The only principle it seems sensible to follow is one of consistency in type of paint: in other words, all opaque or all transparent; all flat enamels, or all permanent oil-based felt-tip pens. If you have time, however, experiment. Paint is easily removed from tin, and trying different combinations is the only way to learn which you like best.

Use a mirror or not as you wish. Tin can be polished almost as brightly. On the other hand, mirrors are not expensive even when cut to order in an odd shape like this one—something like seventy-five cents. Just take along a paper pattern for your mirror man to go by.

A pair of these tulip sconces would be charming in a tiled garden room.

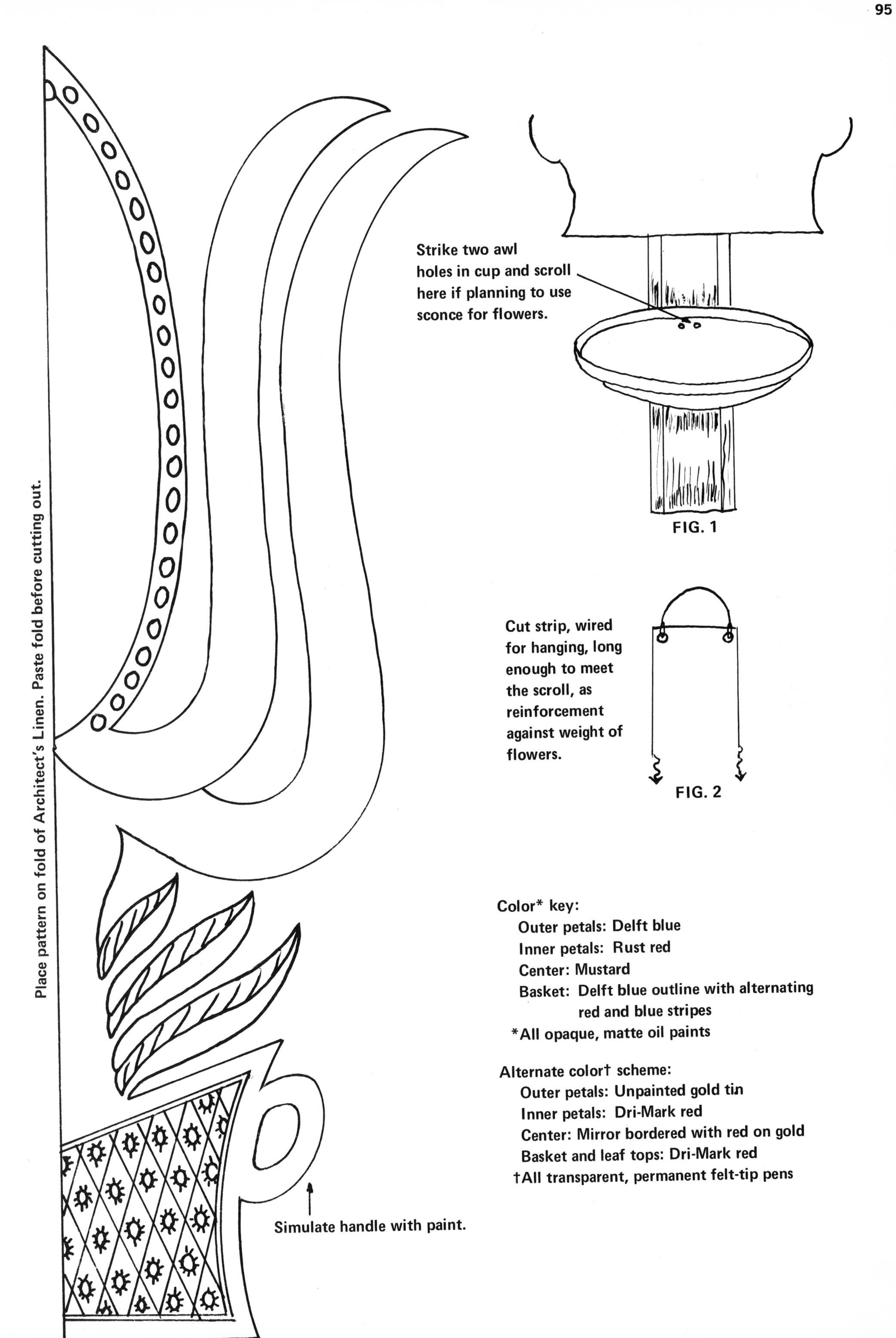

Color* key:
Outer petals: Delft blue
Inner petals: Rust red
Center: Mustard
Basket: Delft blue outline with alternating red and blue stripes
*All opaque, matte oil paints

Alternate color† scheme:
Outer petals: Unpainted gold tin
Inner petals: Dri-Mark red
Center: Mirror bordered with red on gold
Basket and leaf tops: Dri-Mark red
†All transparent, permanent felt-tip pens

PINEAPPLE SCONCE

Can you think of a better use for a ham can?

TOOLS

Snips
Ball-pein hammer and board
Awl
Cold chisel *(optional)*
Long-nosed pliers
Pipe or broomstick

MATERIALS

12-pound Krakus Polish Ham can, including both top and bottom lids
Steel wool
Whole end of soup can
Architect's Linen, scissors, pencil
Rubber cement
Weldit Cement
Upholsterer's tack
Florist wire
Paper clips
Mirror *(optional)*

The lids on this can may be recessed too deeply for your particular can opener, so if you have difficulty, ask your grocer if he will be kind enough to dismember it for you. He will have a commercial opener that cuts efficiently and neatly so that you can use the rims for a mobile, as well as the lids and sides for this sconce.

Hammer both lids and the sides of the can until flat. Do not bother to trim the edges of the lids; they look hand-crafted and pineapple-y just as they are. Burnish them bright with steel wool.

THE FACE OF THE PINEAPPLE

Select the lid with the fewest numbers on it and extract the center portion of it by cutting carefully along the fourth line in from the edge, as indicated in the diagram. Hammer the edges smooth.

Look at the pieces critically for a moment. Would you prefer a mirror there? Or a pineapple? If a mirror, you would then only ornament the border. If a pineapple, you will strike a crosshatch with the cold chisel from the top (Fig. 1), then beat *within the crosshatch,* from the *back,* with the round end of the ball-pein hammer on a flat board. As you can tell from the mathematical inaccuracies in my drawing, I do all this "by eye." My cold chisel is just the width of that top, *angled* line, and that decided for me where to place it. If you don't have such a chisel, you can use an awl and ruler instead. Don't be tempted to make smaller diamonds. I've tried it and find the bolder ones far more effective.

Hammering within the crosshatch from the back will curl the pineapple up, so restrike the crosshatch from the front to flatten it.

Careful to strike exactly the same line!

Ornament the border by making short, curved cuts all the way around. Hammer them flat. Return from the far side, cutting easy curves down to the original cuts as shown. When you get to the top of the pineapple, turn the piece over and repeat the process. (If you are left-handed you would do all this in reverse.)

Note the instructions on the diagram for finishing the pineapple at the top.

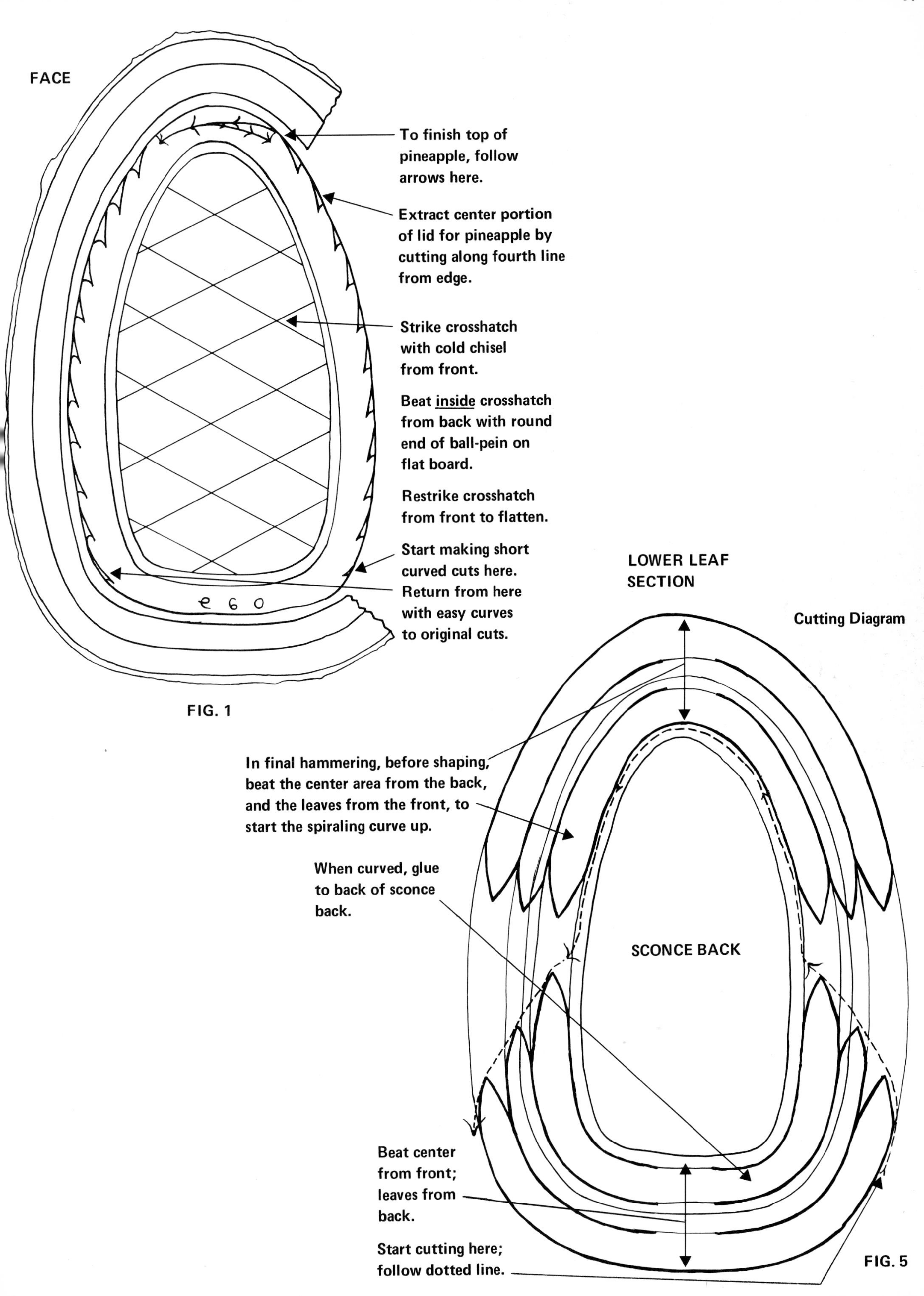
FACE
To finish top of pineapple, follow arrows here.
Extract center portion of lid for pineapple by cutting along fourth line from edge.
Strike crosshatch with cold chisel from front.
Beat inside crosshatch from back with round end of ball-pein on flat board.
Restrike crosshatch from front to flatten.
Start making short curved cuts here.
Return from here with easy curves to original cuts.
LOWER LEAF SECTION
Cutting Diagram
In final hammering, before shaping, beat the center area from the back, and the leaves from the front, to start the spiraling curve up.
When curved, glue to back of sconce back.
SCONCE BACK
Beat center from front; leaves from back.
Start cutting here; follow dotted line.

FIG. 1

FIG. 5

THE SCONCE BACK

Before you start with the second lid, study Fig. 1 and the photograph. Note the way the design takes advantage of the lines stamped in the lid. That explains the simple patterns in Fig. 2; they are mere outlines to make sure you get proper proportions. You might make the leaves ¼" longer than they are here, but any longer than that will seem loose and uncontrolled.

You will notice that the sconce back pattern (Fig. 2) provides only for the leaves and not the pineapple. That is to save precious Architect's Linen. You are, of course, meant to cut up and around the pineapple top, as indicated by the dotted lines in the diagram.

Apply the patterns with rubber cement to the lid. Cut out (see Cutting Out Patterned Designs, page 16, and the diagram); remove the pattern and the glue.

Cut along the proper lines to delineate the leaves. Improve the silhouette of the leaf tips, cutting long, easy curves, and rounding the points the tiniest bit.

Hammer the leaves till limber. Encourage them to curve in the direction you want them to go (see Cutting Diagram, Fig. 5) by hammering from the front or the back as the situation calls for. The metal comes *up* under the hammer, you remember.

Because you are gluing rather than soldering this sconce, you will have to do all the shaping of the leaves before you assemble it. The leaves on the sconce back aren't so difficult. Rub each one back over your forefingers with your thumb, to put a smooth curve in it. Then, holding the base of the leaf with your helping hand, twist the top of it in a spiraling sort of motion so that it is turned toward you, *edge on.* You can see from the photograph that it's only a slight turn and quite easy.

But the lower leaf section requires much more manipulation, because the leaves not only reverse their direction completely, but spiral off to right and left besides. You will have to rub them first *straight back over your forefingers*—way over until they almost form a circle—*with your thumbs;* and then pull them out to the side in a spiraling motion. You can see from the photograph how *their curves* must continue the line established by the *leaves in the sconce back.* By themselves, when properly curved, they resemble a trio of long-horned cows.

When you think you've got the lower leaf section right, put the sconce back down on the floor; slip the lower leaf section into position under it; pin them down firmly with your helping hand; and work with the leaves until they are all "flowing" together.

THE PLUME

Trace the pattern onto a fold of Architect's Linen and *paste the fold together before cutting it out.* Work in a good light and cut carefully so the curves will be graceful. Open up and apply with rubber cement to the burnished side of the can and cut out (Fig. 3).

Strike the awl holes for hanging from both sides and hammer flat. Hammer the plume from the top till it curves around a little.

The second section of the plume is a short strip, 1½" x 4" (Fig. 4). You don't really need a pattern for this. Cut as indicated in the diagram. Hammer well. Rub the leaves over your forefingers with your thumbs, etcetera, as you did with the others. The two outer ones, you notice, spiral off to the side.

Scratch the base of the second section *on the back* and the corresponding area *on the front* of the plume, and glue them together with Weldit Cement. Set aside to dry while you make the scroll.

THE SCROLL

From the side of the ham can, cut a strip 11½" x 1½" (Fig. 6). Draw horizontal lines across the strip to define the areas for leaves, etcetera. You won't need a pattern for this; you've been through all the processes before and the diagram makes them clear.

The only thing that might be new to you is the folding of the scroll-edge (see Scrolls and Candle Cups, page 17). *Pare down* the strip below the leaf heading, ⅛" on each side, cutting along the dotted line.

When you have finished folding the edge, notch the end as shown (Fig. 6), and curve the scroll around a short length of pipe (or broomstick) by tapping it with the hammer. Look at Fig. 7 and Scrolls and Candle Cups (page 17) also for forming the candle cup and Colonial socket.

ASSEMBLE THE SCONCE

The sconce is held together by glue, wire, and the upholsterer's tack.

Holding the pineapple face in place on the sconce back (it will cover the *top* row of leaves), try the plume in position. When you have it right, mark the place on the back of the sconce. Lay the face aside.

Put the plume in place again and scratch through the awl holes to mark their position on the sconce back. Lay the sconce back *face down* on the corner of your hammering board and strike the awl holes.

Scratch the areas of contact where you will be

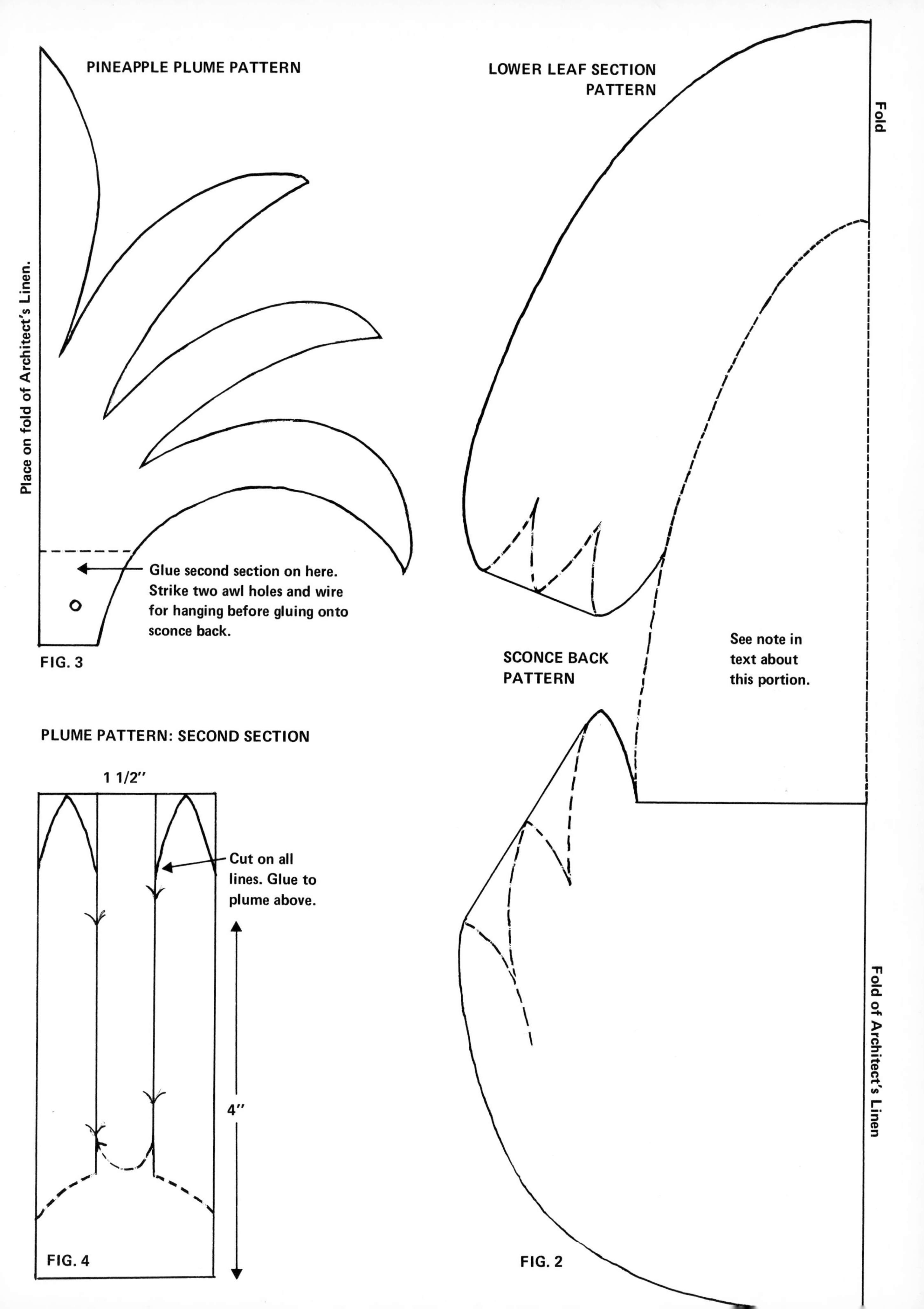
PINEAPPLE PLUME PATTERN
Place on fold of Architect's Linen.
Glue second section on here. Strike two awl holes and wire for hanging before gluing onto sconce back.
FIG. 3
LOWER LEAF SECTION PATTERN
Fold
SCONCE BACK PATTERN
See note in text about this portion.
Fold of Architect's Linen
FIG. 2
PLUME PATTERN: SECOND SECTION
1 1/2"
Cut on all lines. Glue to plume above.
4"
FIG. 4

SCROLL

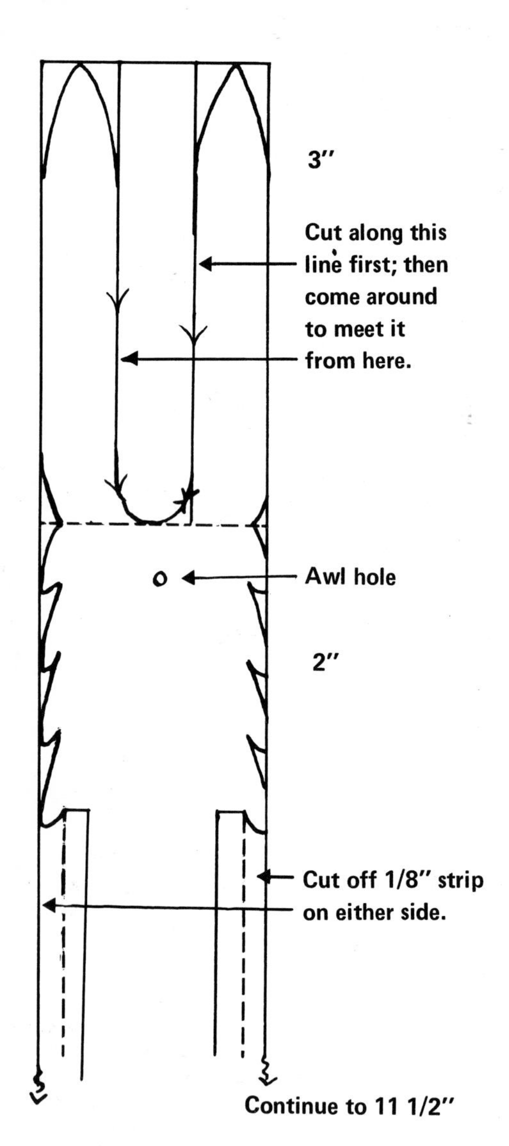

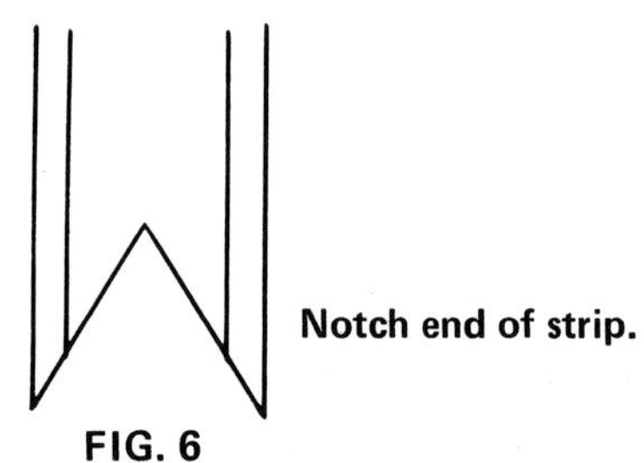

FIG. 6

COLONIAL SOCKET

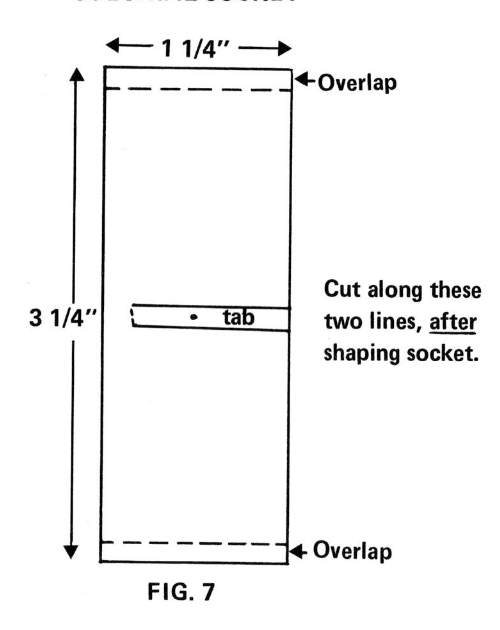

FIG. 7

Hammer strip for candle socket until it curves up; then bend around pipe or broomstick. Glue or solder overlap. Cut central slot and bend tab inside socket and glue down to saucer (or candle cup).

putting glue on the front of the plume and the back of the sconce. Apply glue and allow to become tacky, cutting a 3½" length of florist wire in the meantime. Attach the wire to the plume; press the two pieces together; and secure with paper clips. Let dry.

Strike an awl hole in the center of the heading on the scroll. Place the scroll in position on the pineapple face and scratch with the awl through the hole to mark the spot for the awl hole. Continue in this fashion on down through the layer of leaves.

Scratch all areas of contact to be glued. Apply Weldit Cement and allow to become tacky. Assemble the layers, push the upholsterer's tack through the awl holes, and bend it over on the back of the sconce with the long-nosed pliers. Set it face down on something small and firm, like a frozen juice can, and hammer the end of the tack down firmly.

Adjust all the sections of the sconce so that they are true; secure them with paper clips; and let dry overnight.

The next morning, glue the candleholder to the scroll, by hanging the sconce on the wall, applying glue to the top of the scroll (and the bottom of the candleholder), and, when tacky, carefully centering the candleholder on the scroll.

A distinctive, personal table setting, the airy Mexican candlesticks and the Continental Cache-pot, here used to frame both wild and cultivated flowers most effectively.

A true example of tincrafted decorations in harmony with nature. At left, a spider plant bursts forth from its hanging basket with a spray of tin stars and butterflies hovering about it. At right, the Pennsylvania Dutch Tulip Sconce becomes a perfect container for rooting the little spider plantlets.

The Celebration Chandelier adds an exciting new hanging dimension to a party setting. Pendant ornaments fore and aft, crowned by the warm glow of lit candles, it's certain to generate gaiety and good feelings for all below it. The Canapear Curtain, created from soup cans, becomes a perfect room divider for both contemporary and traditional interiors.

FEATHERY SCONCE

TOOLS

Coarsely-serrated snips
Ball-pein hammer and board
Awl
Oval die or sawed-off screwdriver
Blunt nail
Chopping bowl

MATERIALS

Large, unribbed silver can, 32″ in circumference
Whole end of soup can
Votive candle
Weldit Cement or Epoxy 220
Ruler
Posey Clay
0000 steel wool
Gloves *(optional)*

The basic design of this Feathery Sconce is really very simple, just a long strip of tin with "feathers" cut down from the top and up from the bottom, with an extra set glued on.

RULE OFF THE FEATHERS

From the side of a large, unribbed silver can, cut a strip of tin 24½″ long by 3¾″ wide.

With ruler and awl, score a line across the strip of tin 8½″ from one end, holding the ruler very firmly and pressing hard on the awl. (Two little dabs of Posey Clay on the bottom of the ruler will help to keep it from slipping as you run the awl along it.) This will be the area for your top feathers.

Score another line across the strip 6″ from the other end for the bottom feathers. The space between is the shaft of the sconce.

Perpendicular to these two lines, mark off five feathers, each ¾″ wide, in both the top and bottom areas (Fig. 1). Do not yet cut along these lines; it is much easier to hammer the border ornament on the shaft without loose feathers flapping on each end.

ORNAMENT THE BORDER

As you can see from looking at Fig. 6, the border consists of one ornamented band ⅛″ wide, flanked by two plain bands 1⁄16″ wide. These lines must be perfectly true, so work in a good light, proceed slowly, and press the ruler down firmly.

Start at the top of the sconce shaft by scoring a second line across the strip ¼″ below the first. That is all you need to do here, because you will be gluing on a second set of feathers with its own ornamented border.

Turn the sconce upside down and score *all* the border lines across the shaft. Then do the same on each side of the shaft.

To ornament the border, start at the lower right-hand side and work *up* the center band, tapping the oval die (or sawed-off screwdriver) with the hammer at regular intervals, leaving room for the dots which you will do afterward. Working up the sconce, instead of down, enables you to see exactly where you are going. Keep your eye on the tin, and the position of the die, rather than on the hammer and the head of the die, or you will tap in the design off-center. As you near the top of the band, do a little calculating on the spacing of the design so that you come out even with the pattern.

When you've finished all the oval-dieing from the front, turn the sconce over and tap the dots between the ovals with the blunt nail.

CUT THE FEATHERS

Once the border design is complete, cut the feathers. Notice that the center feather is ½″ shorter than the others.

Draw a light line across both sets of feathers 1″ from their ends. Cut curved tips down to that line.

Finally, make the diagonal cuts ⅛″ apart, starting at the tip of the feathers and working down to within ½″ from the shaft border (Fig. 3).

THE SECOND SET OF FEATHERS

Cut a second piece of tin 6¼″ x 3¾″ (Fig. 2).

Across one end, score the border design (Fig. 6).

Cutting *down to* that border, remove a ⅜″ strip from either side.

Mark off the five ⅝″ feathers, cut like the others, and hammer flat.

Glue this extra set of feather in place on the sconce with Weldit Cement or Epoxy 220. Secure with paper clips till dry.

THE CANDLEHOLDER

Beat the whole end of a soup can in your kitchen chopping bowl with the round end of the ball-pein hammer until it conforms to the shape of the bowl.

Scratch it on the underside with the awl to roughen it, and glue it to the top of the lower center feather about 2″ from the border.

Reinforce the center feather by gluing a strip of tin 6″ x ¾″ on the center back of the sconce, allowing it to extend 2″ below the shaft of the sconce under the center feather. Glue it only to the sconce back, not to the feather.

To curl the feathers, pinch them between your thumb and forefinger, starting at the base of each and pinch-pinch-pinching all the way to the tip to make them curve toward you.

To make the feathers fan out, grasp the very tip with your fingers, and with a spiral twist of your wrist, pull it out to the side. Pinch it some more. Spiral it again. Study the photograph. Your eye and your instincts will guide you.

Suspend the sconce either from an awl hole in the sconce back or by a wired tab glued to the back.

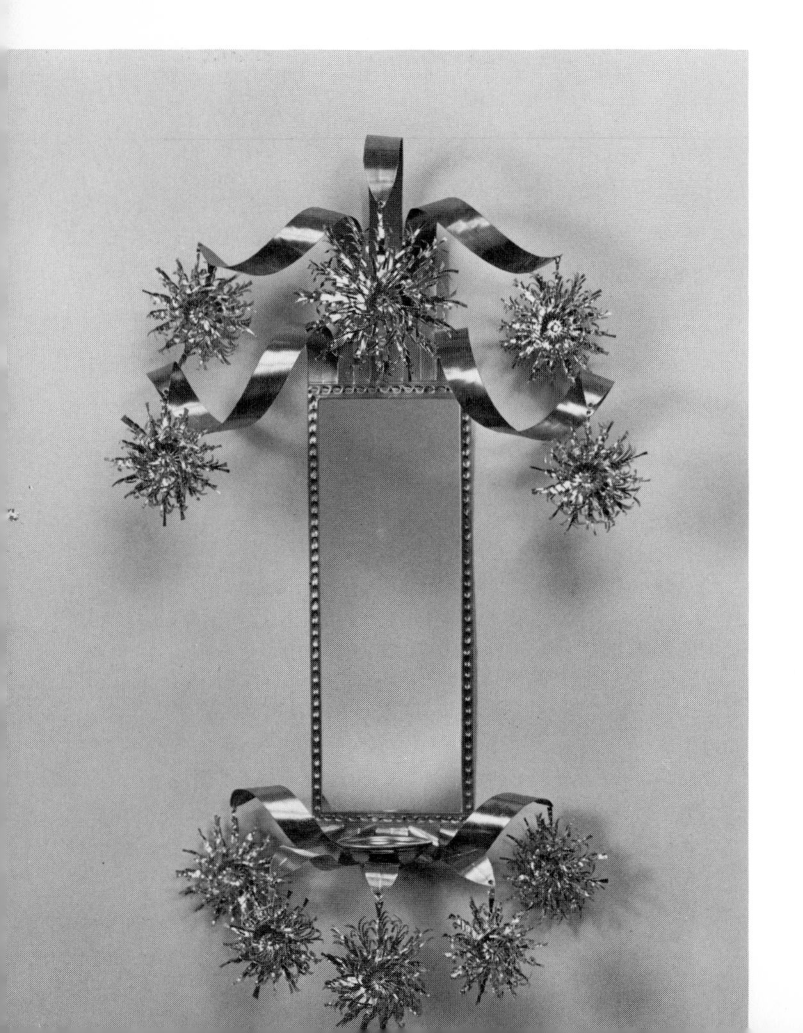

FROSTY STAR SCONCE

Although this sconce is almost identical in design to the preceding Feathery Sconce, the two are as different in mood as Christmas and Easter—mostly on account of those fabulous Frosty Stars.

TOOLS

Coarsely-serrated snips
Ball-pein hammer and board
Awl
Long-nosed pliers
Oval die or screwdriver
Blunt nail
Kitchen chopping bowl

MATERIALS

Large, unribbed silver can at least 24″ in circumference
10 lids from 1-pound coffee cans
Whole end of soup can for candle cup
Mirror 3¼″ x 9¼″ *(optional)*
Weldit Cement
10 sew-on-type rhinestones
Florist wire
Ruler and Posey Clay
Frozen juice cans or dowels for curving arms
#2 knitting needle for links

FEATHERY AND FROSTY STAR SCONCES

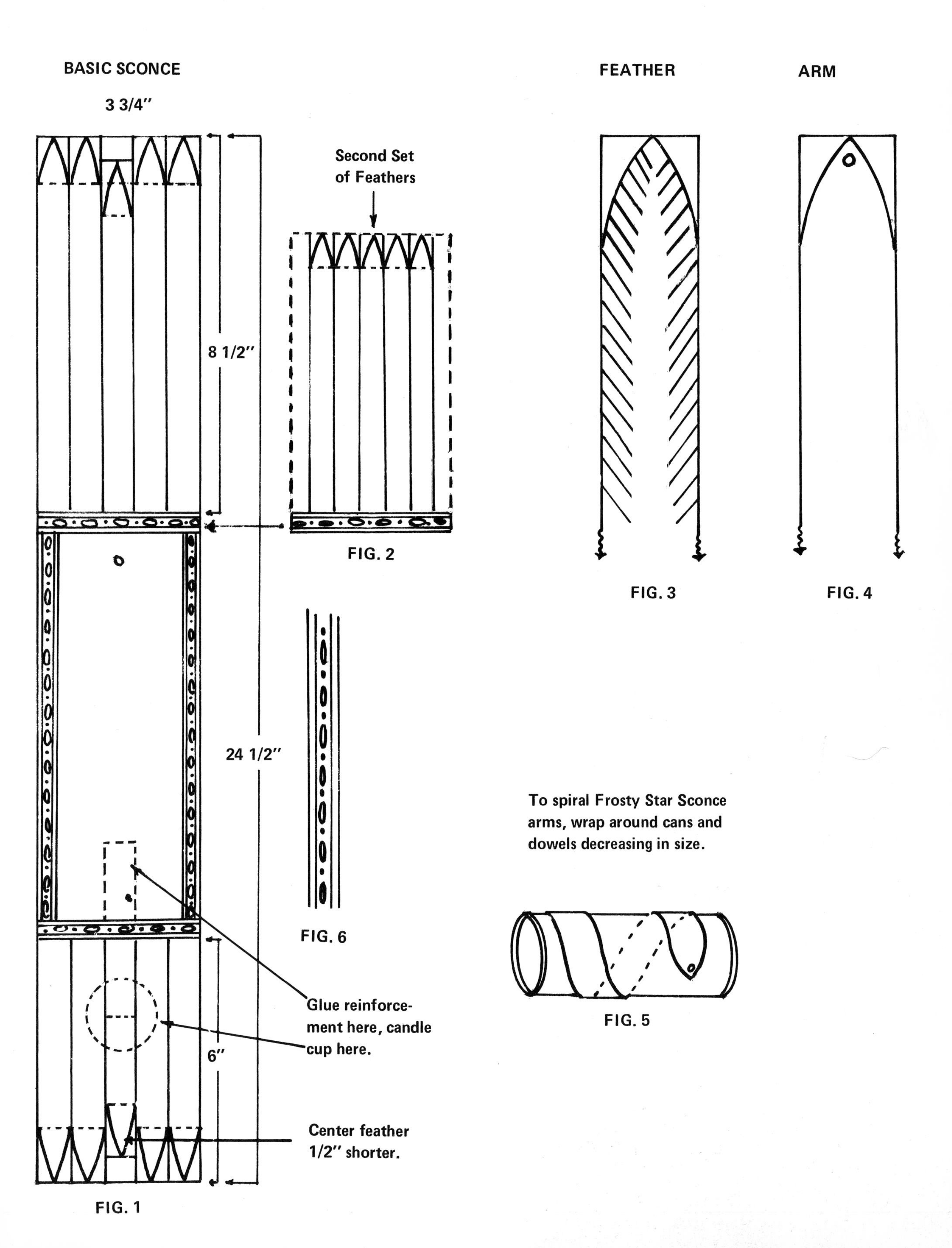

FIG. 1

FIG. 2

FIG. 3

FIG. 4

FIG. 5

FIG. 6

From the sides of a large, unribbed silver can, carefully cut a strip of unblemished tin measuring 24½″ x 3¾″.

Score a line across the tin 8½″ from one end with ruler and awl. Two little dabs of Posey Clay on the bottom of the ruler will help to keep it from slipping as you run the awl along it.

Score another line across the tin 6″ from the other end.

Define and ornament the border, referring to the instructions given in the Feathery Sconce preceding. Because you will *not* be adding a second set of arms to this sconce, ornament the border at the top as on the sides.

When the ornamentation is complete, cut the arms, noting that the middle arm is ½″ shorter than the others.

Draw a light line across both sets of arms 1″ from their ends. Cut curved tips down to that line.

Strike the awl holes in the tips of the arms, first from the front, then from the back; hammer flat from the back (Fig. 4).

Spiral the arms by curving them forward over cans or dowels, starting with something the size of a large frozen juice can, and gradually working down to a dowel the size of a broomstick, and winding them out the side simultaneously as indicated in Fig 5.

If you inadvertently get a "bend" in an arm, hammer it flat and proceed as before.

THE CANDLEHOLDER

The candleholder is made from the whole end of a soup can, beaten in your kitchen chopping bowl with the round end of the ball-pein hammer until it conforms to the bowl. Glue it in place on the lower center arm with Weldit Cement, scoring the areas that touch.

You will need a hook for hanging on the back—just a simple strip of tin wired with a loop. Score and glue.

Glue the mirror on also, unless you find the sconce quite shiny enough as it is.

THE FROSTY STARS

Nothing remains to be done but the ten Frosty Stars, the diagram for which you will find further on in the book with the Frosty Star Mirror, page 120.

You will notice from the photograph of the sconce that the stars are graded in size. The large star top center is made from a whole 1-pound coffee can lid. The four on either side of it are ¼″ smaller and require that you cut down the lids to the *second* concentric circle. At the base of the sconce, the two outermost stars *and* the one in the center require you to cut down the lids to the *third* concentric circle; and the two remaining stars should be cut down to the *innermost* concentric circle.

Hammer the lids flat.

From a piece of gold-lacquered scrap tin, cut ten fringed circles about the size of a nickel. Score and glue them to the center of each lid. Glue a sew-on-type rhinestone to the center of each fringed circle. Let dry overnight before cutting the stars.

When cutting each star, make sure to leave the tip of one section *wide* enough to acommodate an awl hole from which to hang the star.

Make ten small links by wrapping florist wire around a #2 knitting needle and cutting through the coil. Link the stars to the sconce with the long-nosed pliers.

DAMASK ROSE SCONCE

When a friend brought me her French tôle sconce for repair, I asked her if she'd mind my making a copy of it to share with those of you who can solder. I have not painted my copy because I like it silver, but the original was done, most ingenuously, in solid unshaded matte oils—a dusty pink rose, forest green leaves, beaver-brown light fixture and rose stems, and a garnet red bow.

TOOLS

Plain-edged snips
Ball-pein hammer
Awl
Long-nosed pliers
Round-nosed pliers
Chisel
Screwdriver
Propane torch and asbestos mat
Vise

MATERIALS

5 lids, grading from 3¼″ to 1½″ in diameter, for the rose
Sides of unribbed, unlacquered cans, for leaves
4 16-gauge straight florist wires, for stems
10-gauge wire for light fixture
Sandpaper
Mechanic's or Mystic tape
Solder, Dunton's Tinner's Fluid
Electrical socket and wire
Candle sleeve
Paints *(optional)*
1 lid for candle cup

THE ROSE

Both the French and the Italians have been making roses like this for centuries.

For the petal layers you will need five lids varying in size from 3¼″ to 1½″ (Fig. 1), which means that you will have to cut down the lid of a small frozen juice can for the smallest one.

Hammer the lids flat. Make an awl hole in the center of each large enough to allow the 16-gauge florist wire to pass through.

Divide each lid into eight equal sections (see Dividing Lids Into Equal Sections, page 15), *except* for the smallest. Divide that into four equal sections.

Trim each section as shown in Fig. 1, referring also to the Apothecary Rose diagram, page 48, for the fast-rounding technique.

Beat each individual petal from the *back* in a beer bottle cap with the *round* end of the ball-pein hammer.

Assemble the layers on a well-sanded 11″ florist wire so that their petals alternate. Clamp the stem of the rose in the vise just under the petals, and solder them in place from the top. Solder them again from the back.

THE LEAVES

Obviously, the designer of this sconce was no horticulturist, because these leaves look more like ivy than roses, but copy them anyway and cut fourteen (Fig. 2). Vein with the chisel from the front. Serrate the edges with the screwdriver from the back. Hammer flat.

THE STEMS

Cut three rose stems from straight 16-gauge florist wires (or brass welding rods, if you prefer), one 14″ long, and two 12½″ long. Sand them well.

If you want the stems to form a solid unit with the scroll (for the light fixture), tin each one individually up to a height of 6″. This will make it easier to solder them all together at the end. But, if you prefer to catch them only at the "waist" with the bow, leave them as they are.

Cut five 4″ lengths of 16-gauge wire for the small side branches and sand well. Bend in half over round-nosed pliers, and squeeze the loops shut as tightly as possible with the long-nosed pliers. Clamp the loops in the vise, and bend the ends out in a pleasing curve, looking to the photograph for guidance. Tin the loops.

Solder the leaves onto all the stem-ends. There will be one remaining, which will go in the center under the rose after the sconce has been assembled.

Solder the branches onto their main stems. Notice that two of these branches are soldered, one under the other, on the central rose stem.

THE BOWKNOT

You will find instructions for this under French Provincial Bowknot, page 92.

Make the tab 2″ long. Should it be overgenerous, clip it off.

THE LIGHT SOCKET OR CANDLE CUP

The sconce had been electrified at some point along the way. Perhaps you might like to know how it was done.

Cut down a lid to a diameter of 2⅜″. Hammer the edges if necessary.

Strike two awl holes in the lid (Fig. 4), one in the exact center, ⅛″ wide, for the tip of the scroll-wire to fit into. Strike another alongside the first, ¼″ wide, for the light cord to pass through, and hammer flat.

Cut a 12″ length of 10-gauge wire for the scroll. Shape it around a broom handle to resemble Fig. 3, placing it on the diagram as you go along to be sure you are getting the curves in the right places. Tin the base of the scroll wire as you did the leaf stems.

From a strip of tin ½″ x 1″, make a sleeve to fit around the top of the scroll wire for the candle cup to rest on (Fig. 3). To shape the sleeve, clamp the scroll wire in the vise; lay the end of the strip against the tip of the wire; and hold the two firmly together with the long-nosed pliers as you wind the strip around the wire with your helping hand. You will, of course, have to move your pliers as you wrap, both to hold what you have already wrapped and to get out of the way of the strip as it comes around.

The sleeve may be loose enough to slide up and down the wire, but no matter. Lower the wire's position in the vise so that the bottom of the sleeve rests on the vise and the top of the sleeve comes to within 1⁄16″ from the tip of the wire (Fig. 3). With torch and acid, clean the tip of the wire and sleeve.

Place the candle cup in position on the sleeve and solder.

The candle socket is made from a strip of tin 1⅛″ x 3¼″, hammered around a length of pipe to overlap ⅛″. This will accommodate the average commercial candle sleeve nicely. Solder the seam. Then solder the socket to the cup.

ASSEMBLE THE SCONCE

Bend the two leaf sprays that will flank the rose *out to the side at a 45-degree angle* at a point 5½″ below their tips, clamping them firmly in the vise to make a neat angle.

Place all three leaf sprays in proper juxtaposition on the asbestos mat, stems flush. Solder the stems together.

Bind the rose stem and scroll wire in position on top of the leaf sprays by wrapping Mechanic's or Mystic tape around the base of the stems.

Put the bowknot in position and wrap the tab ends around the stems. Solder all together from the back.

Solder on the loop for hanging (Fig. 5) up behind the rose.

Solder on the remaining leaf in the center under the rose.

Remove the Mechanic's tape; clamp the stems together in the vise; and flow heat along them to "set" them together.

Rinse off and polish the sconce before painting.

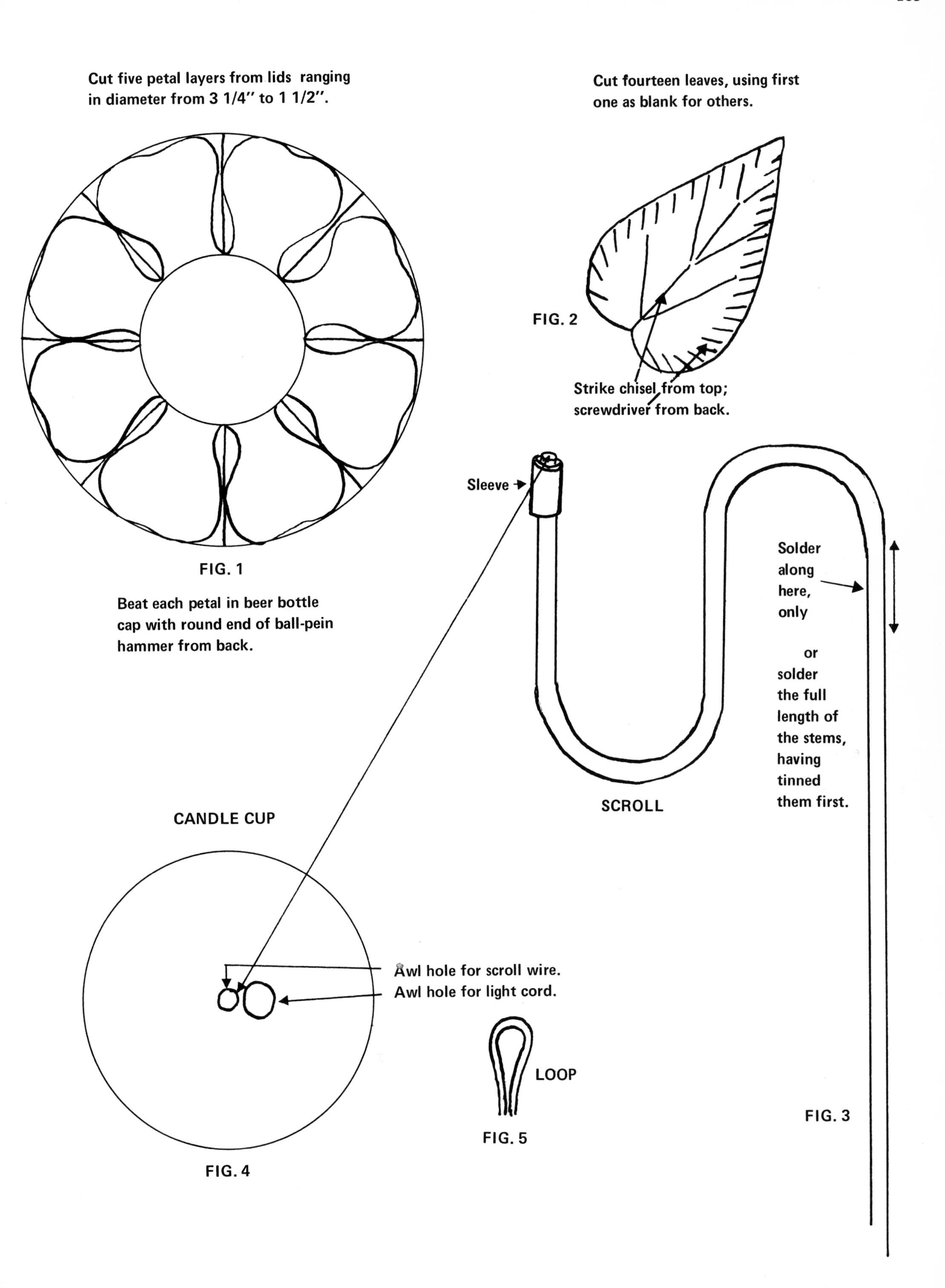
Cut five petal layers from lids ranging in diameter from 3 1/4" to 1 1/2".
Cut fourteen leaves, using first one as blank for others.
FIG. 2
Strike chisel from top; screwdriver from back.
FIG. 1
Beat each petal in beer bottle cap with round end of ball-pein hammer from back.
Sleeve
Solder along here, only
or solder the full length of the stems, having tinned them first.
SCROLL
CANDLE CUP
Awl hole for scroll wire.
Awl hole for light cord.
LOOP
FIG. 5
FIG. 4
FIG. 3

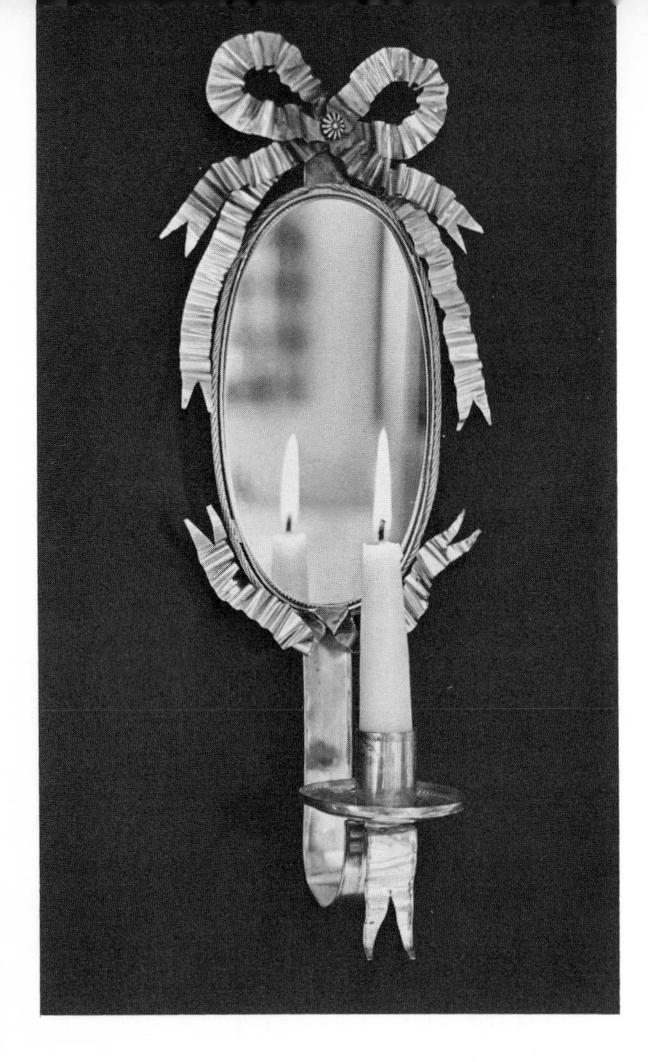

LOUIS XVI MIRRORED SCONCE

Elegant and easy, everything considered, but you do have to know how to solder, and teasing that bow into shape requires a sort of instinct. Worth attempting, though, don't you think?

TOOLS

Plain-edged snips
Round-nosed pliers
Ball-pein hammer and board
Propane torch
Asbestos mat
Vise

MATERIALS

3-pound coffee can stripped of lithographing
Marshall's Kippered Herring can
Mirror
Sash cord wire
Florist wire
Weldit Cement
Solder, Dunton's Tinner's Fluid, baking soda
Upholsterer's tack
Catch basin
Krylon Crystal Clear Acrylic Spray

THE SCROLL

The scroll for this sconce should be 16″ long and run up the entire length of the sconce back, extending 1½″ above, in order to hold the bowknot.Use a large, all-silver can for this (Idahoan Mashed Potatoes come in them, among other things) or a 3-pound coffee can stripped of lithographing. The ribbons will be cut from it also. Refer to the section on Scrolls and Candle Cups (page 17), *noting that the base of this scroll repeats the ribbon motif* in the bowknot.

THE MIRROR FRAME

The mirror frame is made from the entire top of the herring can, and the only trick is to get it off, rim and all, without dumping the contents all over the floor. You will need to put a catch basin under the can opener for the juice to spill into. Place the can under the wheel of the opener *sidewise* and take off the whole lid. Drippy business, but worth the nuisance, it's such a lovely oval.

Since you will be gluing a mirror on top, you need not remove the lithographing unless you want to, but do check to see that the back is clean and smooth.

Turn the lid face down on the asbestos mat; bend the sash cord wire to fit around it, meeting at the base; and solder the wire to the lid in six places (Fig. 1).

Solder the scroll in place also.

CRIMP THE BOW

Why it is that crumpled ribbon looks so gay and festive when made of wood or metal, but entirely unacceptable if made of velvet or silk, I don't know; unless, perhaps, the carving and crumpling of rigid materials makes them seem as soft as velvet or silk. Anyhow, that has ever been the fashion of a fussy French bowknot, and it looks quite right.

Including the ribbon at the base of the mirror, the bow consists of four ½″ wide strips of tin, one measuring 14″ in length, two measuring 12″, and one measuring 8″.

Round the ends of the 14″ strip and give it a waspwaist (Fig. 2). Strike awl holes at both ends and in the middle. Insert the stem of the upholsterer's tack into the *center* hole; bring the bow-ends around and hook them onto it; hammer the bow onto your hammering board (Fig. 3).

Crimp the loop with the round-nosed pliers, starting at the center, twisting your wrist to right and left, making hills and valleys, and moving the pliers from a valley into a hill and down onto the next valley, etcetera, until you have that errant loop facing you squarely. One thing you don't need to worry about is overdoing it: the more you twist and torment that strip of tin, the more like ribbon it will look.

Fold the three strips for the ribbon ends as shown in Fig. 4. Cut notches in their tips, and crimp them, working from the tips back to the fold. (You will hold them in the air to do the crimping, and tack them to the board to put the *lateral* curves in them.)

Assemble the bow and its two sets of ribbon ends (one 8″ and one 12″) by inserting the upholsterer's tack through their awl holes and soldering them all together from the back.

Strike an awl hole in the top of the scroll where you will attach the bow. See how far the tack protrudes through the scroll and nip it off so that it will be flush when in position. Assemble all together and solder from the back. Tease the longer set of ribbons to make them frame the mirror nicely, and lightly solder them to it from the back.

As for the other 12″ ribbon, solder it to the base of the frame from the *front.* Fold the two ends back on themselves at an angle so that they conform to the frame nicely.

Be sure to rinse the frame in water with a little baking soda, so there will be no acid left to eat it. Dry well, and spray the frame twice over, lightly, with Krylon Crystal Clear Acrylic, then glue the mirror in place.

It's quite lovely, don't you think? Good enough to take a little pride in.

LOUIS XVI MIRRORED SCONCE

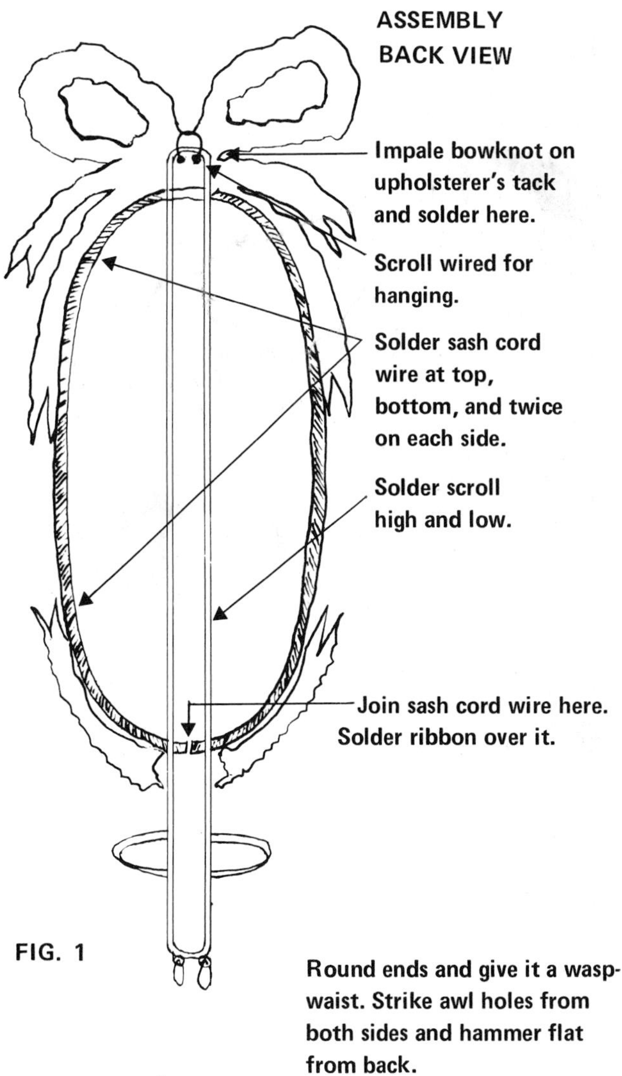

FIG. 1

Round ends and give it a wasp-waist. Strike awl holes from both sides and hammer flat from back.

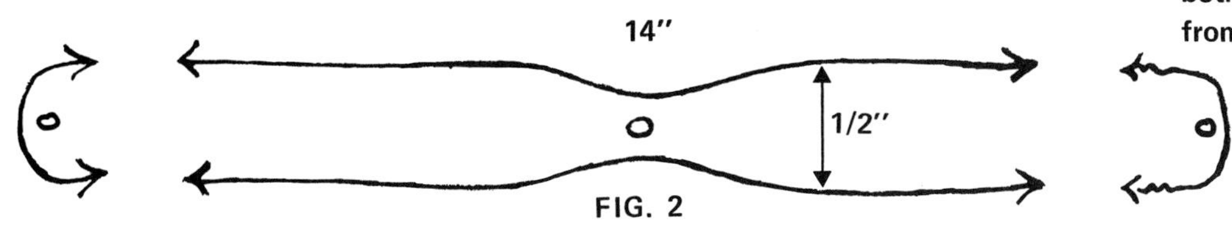

FIG. 2

Insert upholsterer's tack in center hole; bring ends around and hook under; hammer onto board to crimp with round-nosed pliers.

FIG. 3

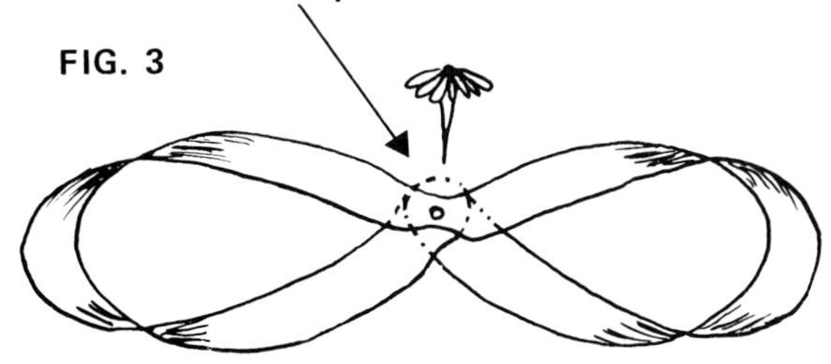

Fold ribbon ends at angle and crimp. One 12″ ribbon and the 8″ ribbon will join the bow above. The other 12″ ribbon, more loosely folded, will frame the base of the mirror.,

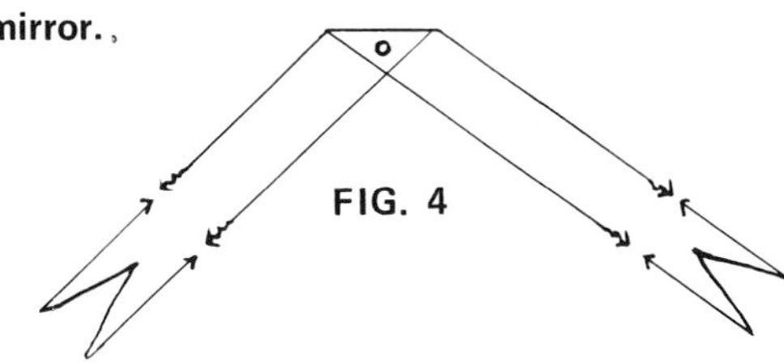

FIG. 4

MINI-MENORAH

No matter how quietly monochromatic you may wish a room to be, you need a few bright splashes of color, and I have found that in my sea-green living room, the clear orange and shocking-pink candles in my two Menorahs provide that pleasing touch. The fact that one of the Menorahs is a mini supplies a dash of humor as well.

TOOLS

Snips
Hammer and board
Long-nosed pliers
Propane torch *(optional)*
Asbestos mat *(optional)*

MATERIALS

Tall, narrow can
½ rim from a small frozen juice can
½ rim from a 1-pound coffee can
1 box birthday candleholders
1 box largest candles to fit
Epoxy 220 or solder, Dunton's Tinner's Fluid
Floratape
Krylon Bright Gold or Silver Spray Paint
Piece of gold-lacquered tin or mirror

The ideal can for a Mini-Menorah is one that is no larger around than a small frozen juice can and half again as tall. You'll have to hunt around on the gourmet shelves to find it. It will probably be an imported item, perhaps Sucroma—a vanilla-flavored sugar—or some other delicacy of European origin where cans are not standardized as they are here and come in all sorts of curious shapes.

LEAF PATTERN

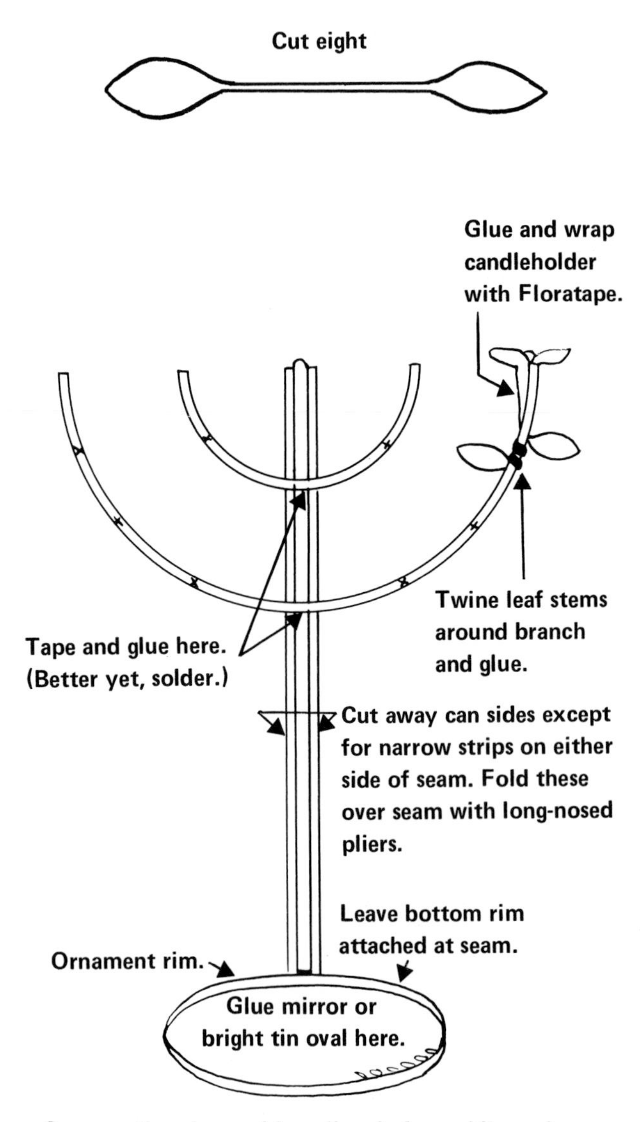

Spray entire piece gold or silver before adding mirror.

The reason for such a skinny can is a simple one of proportion: you will be using the bottom rim of the can as the base of the candelabrum, and obviously, you won't want it to be as broad as the arms. The diagram indicates the general proportions.

Assuming that you have found the ideal can, first remove the bottom lid, then the top rim (see Removing Rims, page 14).

Next, *almost* remove the bottom rim, cutting from one side of the seam around *to* the other side. Then, cut away the can sides as indicated in the diagram, leaving approximately ⅛″ strips on either side of the seam. Fold these over the seam with the long-nosed pliers. If you can solder, apply heat along the seam until the solder in it flows forth and binds them all together.

Ornament the rim (see Ornamenting Rims, page 15) and squeeze it into an oval with your hand.

ATTACH THE ARMS

Cut the *rims* from the small frozen juice can and 1-pound coffee can *in half*, and glue them in position on the seam with Epoxy 220. (Better yet, solder them on.) Stretch the longer arms apart *sidewise* a little to encircle the smaller ones perfectly.

BIND ON THE LEAVES AND CANDLEHOLDERS

Unless you think you can cut them out freehand, trace and apply the leaf pattern with rubber cement to the side of the can. Cut eight of them; hammer them flat; and wrap them around the arms where indicated. Glue, or solder, in place.

To attach the candleholders, apply Epoxy 220 to the *inside* [illegible] of each of the arms, and to *one* side [illegible] each of the candleholders. Press together [illegible] Floratape.

PAINT THE MENORAH

Spray-paint the whole Menorah with Krylon Bright Gold or Silver Paint. If gold, cut a piece of matching gold-lacquered tin to fit on the bottom of the oval base and glue on. If silver, have your local glass company cut you a mirror to fit.

MEXICAN CANDLESTICKS

They look so simple, don't they? Take another look at them in the color photograph on page 101. Innocent little airy fellows. But the fact is they are the hardest thing in the book. You really have to know how to solder, and even then, you'll find it difficult to add the cone at the top without unsoldering the joining of the wires just below. If you're not careful, the heat will flow right on down and undo your previous handiwork. However, that's the worst of it, and isn't it a plus to have them hold flowers as well as candles?

TOOLS

Plain-edged snips
Round-nosed pliers
Long-nosed pliers
Ball-pein hammer and board
Screwdriver
Short length of pipe or broomstick
Propane torch
Asbestos mat
Vise

MATERIALS

Deviled ham can, 3″ in diameter
Lid from 1-pound coffee can
Architect's Linen, scissors, pencil
Rubber cement
14-gauge wire
Sandpaper
Scrap tin for discs
Weldit Cement *(optional)*
Solder, Dunton's Tinner's Fluid
Spool of thread for a jig
Krylon Gloss White Enamel Spray Paint

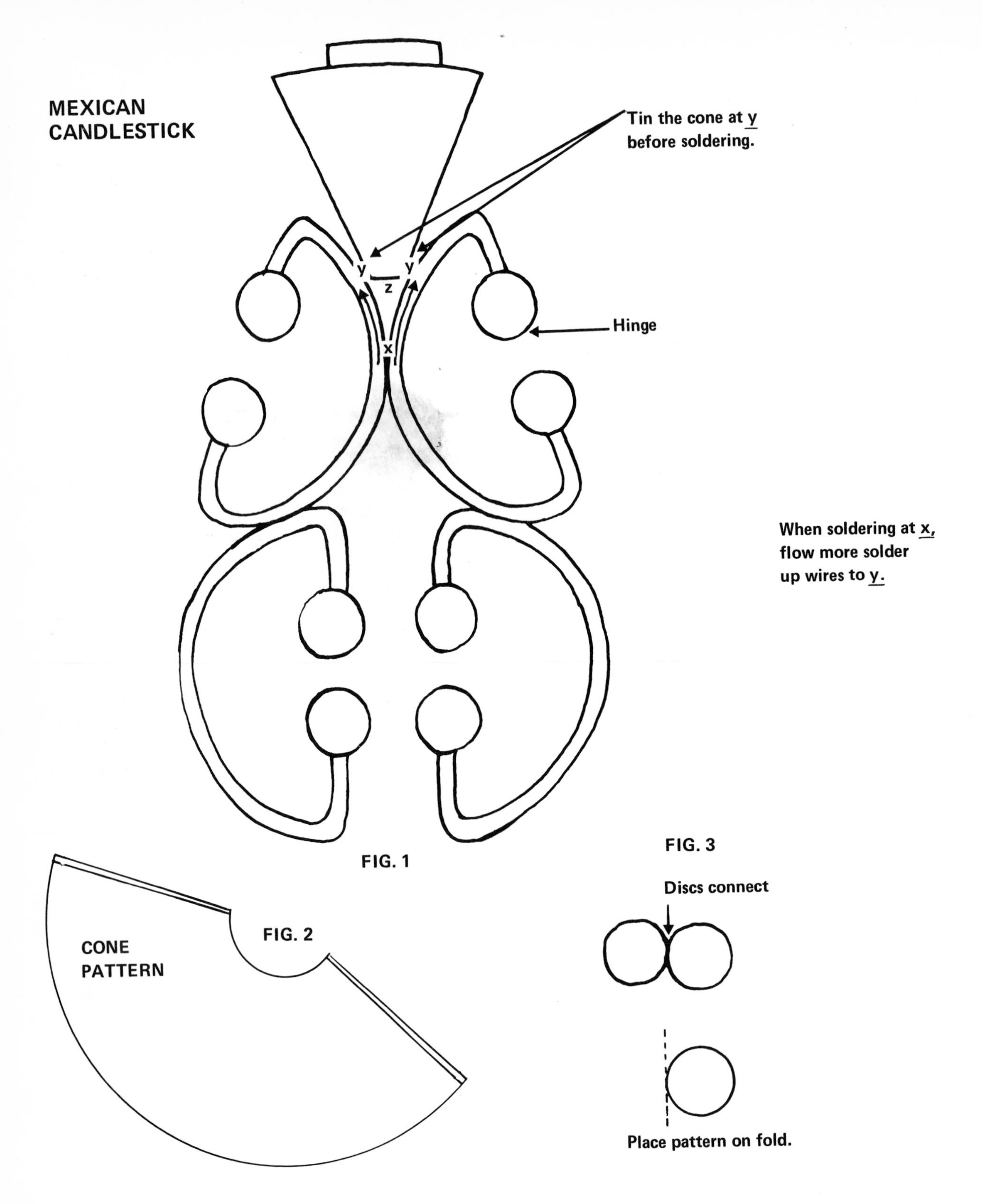

THE SHAFT

For one candlestick, clean a 20″ length of 14-gauge wire by rubbing it well with sandpaper.

From it, cut two 5″ lengths and two 4½″ lengths. Shape these around the ham can, which you will use, eventually, as the base of the candlestick. Remove the four shaped half-circles.

With the long-nosed pliers, bend over the ends (Fig. 1). Make sure each pair matches up. The smaller pair goes on top.

If you are going to glue on the little discs, wait till later, but if you are going to solder them on, do it now before assembling the wire crescents. Cut out the disc pattern on a fold of Architect's Linen (Fig. 2). Apply with rubber cement to a scrap piece of plain tin. Cut out eight hinged discs.

Hammer them flat. Improve the outline. Be careful not to cut through the hinge; it helps to have that tiny joining.

Cup them a little by beating on a flat board with the round end of the ball-pein hammer.

Fold them over the tips of the wires. Hold them in position in the vise and solder.

Assemble the four wire crescents by laying them flat on the asbestos mat in proper juxtaposition. Solder in place. When soldering at *x*, flow extra solder on up to *y* where the candle socket will later join it. This will tin the wire and make it easier to attach the cone.

THE CONE

To make the cone, hammer the lid of a 1-pound coffee can until it is as flat as you can get it. You will notice that it is already curving up, cone-fashion.

Cut it according to Fig. 2 and shape it into a cone either by rubbing it round and round with your thumbs and fingers, or by bending it around a small funnel, or by working it with the round-nosed pliers until it curves around and stays in position of its own accord with a 1⁄16″ overlap. Solder the seam.

For the candle socket, cut a piece of tin 1″ x 3⅛″. Hammer it until it curves up, and shape it around a short length of pipe or broomstick (see Scrolls and Candle Cups, page 14). Squeeze it past the meeting point until it remains in position with ⅛″ overlap. Pinch any unevenness in the cylinder with the long-nosed pliers to make it a perfect tube. Solder the seam.

To solder the socket inside the cone, set the cone in the center of an empty spool of thread to hold it upright. The socket, if nicely rounded, should sit neatly in the cone and make an easy connection.

ASSEMBLE THE CANDLESTICK

Before you solder the cone onto the shaft, tin the two spots on its sides where it will touch the shaft. Then, moving the asbestos mat forward under the vise, as shown in the photograph on page 101, place the cone upside down on the mat alongside the vise. Secure the shaft of the candlestick upside down in the vise so that it makes proper contact with the cone. Slip a sliver of pounded solder between the shaft and the cone, if you think you need it, and apply heat until it melts.

The worst is now over. Solder the shaft to the ham can, holding the shaft in the vise and slipping the can into position under it.

Now is the time to glue the discs in place if you haven't already soldered them on. Daub their inner surfaces, and the tips of the wires, with Weldit Cement. When tacky, fold them carefully into position. Let dry before spraying with Krylon Gloss White Enamel.

CELEBRATION CHANDELIER

This is pictured in color on page 102. Smashing, don't you think? A real fun party thing, and easy!

TOOLS

Snips
Hammer and board
Awl
Compass
Round-nosed pliers
Long-nosed pliers

MATERIALS

Circular 4-gallon can, measuring 10″ in width and 13″ in height
16 small frozen juice cans
16 lids from small frozen juice cans
Shelf paper
Rubber cement
Epoxy 220 or Weldit Cement
Krylon Flat White Spray Enamel
Krylon Bright Gold Spray Enamel
5′ decorator chandelier chain
16 3″ candles
Balsam wreath
Imperial Star

Except for the round 4-gallon can, which is essential, the materials listed here are suggestions only. Every party has its own special theme, and yours will undoubtedly call for different colors and ornaments, or perhaps no ornaments at all but more candles instead. In any case, the way I have made mine will give you something to go by.

SUN AND SPARKLER STAR PENDANT

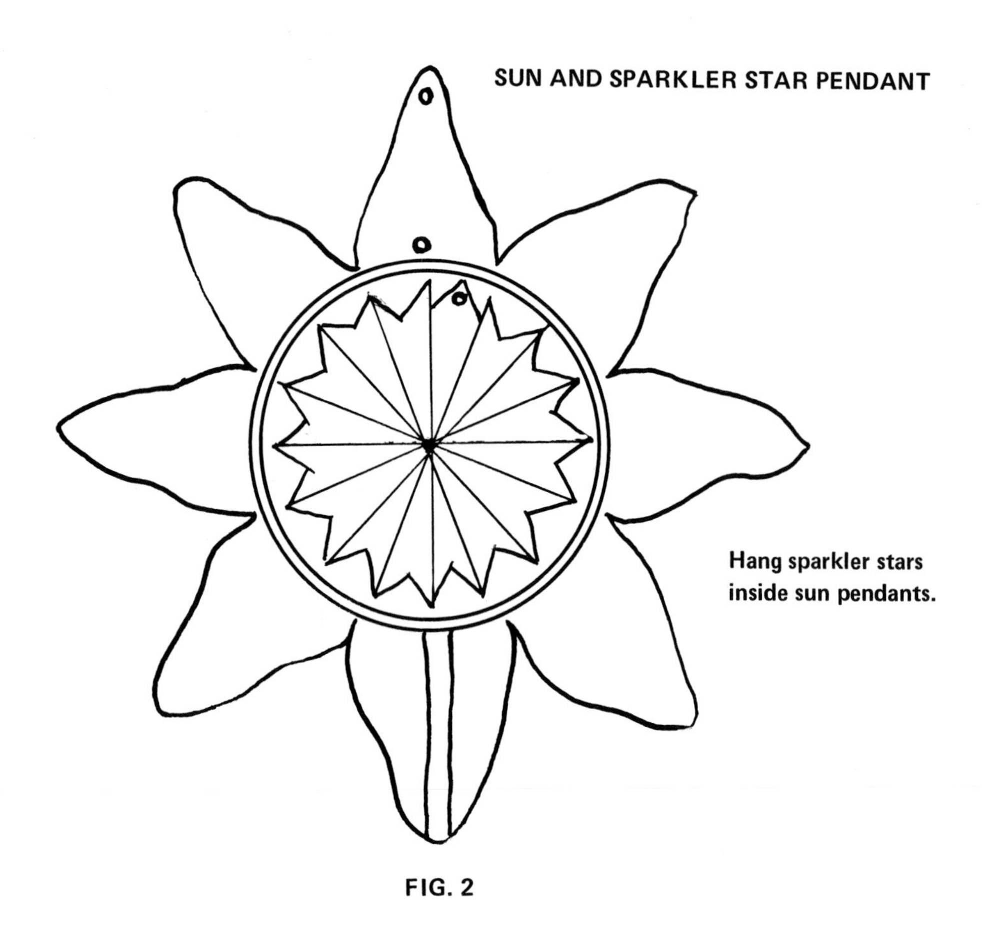

Hang sparkler stars inside sun pendants.

FIG. 2

Round arm-ends either way.

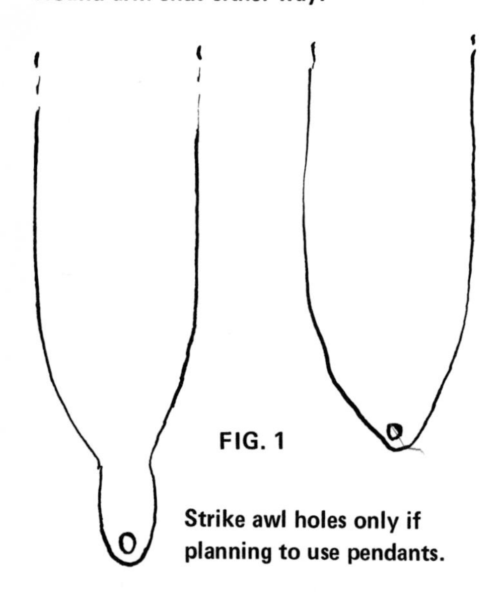

FIG. 1

Strike awl holes only if planning to use pendants.

Strike three pairs of awl holes in end of can for wiring chain to chandelier.

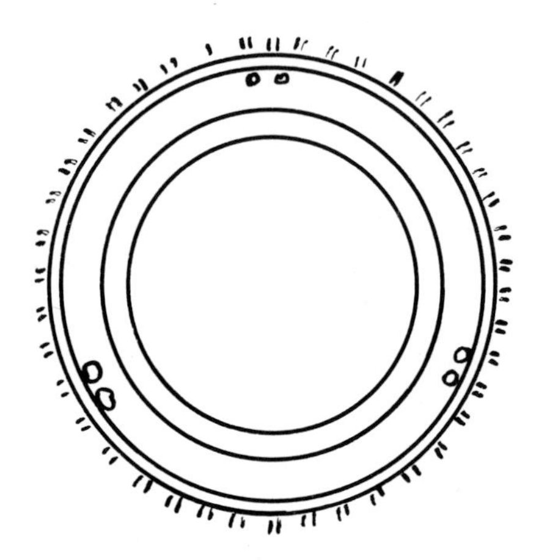

FIG. 3

THE BASIC FRAME

If you feel the chandelier must match something in particular which requires its being painted with a brush, do it now when the can is still in one piece. But if the chandelier is to be spray-painted a solid color, wait until later.

With the compass, mark off a line around the side of the can 3″ from the top rim (see Cutting Down Cans, page 14). Every other arm on the chandelier will be cut down to that line.

Remove the top rim of the can (see Removing Rims, page 14).

Divide the sides of the can into 32 equal sections (see Dividing Cans Into Equal Sections, page 15), using shelf paper and rubber cement, *not* newspaper, which absorbs the cement and sticks to the can.

After cutting along the creases, pull the arms out horizontally in an easy curve. (A tall round stool is ideal to set the chandelier on while you work on it.)

Cut off every other arm at the compass line. Round all arm-ends as you wish (Fig. 1).

If you are going to use pendants, strike the awl holes in the arm-ends now.

Also, strike three pairs of awl holes, from which to suspend the chandelier, in the bottom of the can next to the rim, as shown in Fig. 3.

THE CANDLEHOLDERS

Crimp the bottom lids from 16 small frozen juice cans (see Scrolls and Candle Cups, page 17). Glue in place on the short arms of the chandelier.

As for candles, the simplest expedient is the small votive type (be sure they are dripless), which sits neatly in the crimped cup without benefit of socket or glue. Since mine was a Christmas chandelier, I used the inexpensive 3″ Hanukkah variety, which resemble the old-fashioned Christmas tree candle. (They looked *so* inexpensive and paraffin-y, in fact, that I spray-painted them gold.) Sockets for these are made with strips of tin ¾″ x 2½″, wound around a dowel and squeezed into perfect cylinders with the long-nosed pliers to overlap ⅛″. Do not bother to glue the overlap; just test each one for size and glue it onto the candle cup.

When dry, spray-paint the chandelier with Krylon Flat White Enamel.

SPARKLER STARS

To make the sparkler stars which hang in the center of the sun pendants, remove the bottom lid from each of the small frozen juice cans. Cut them down 1⁄16″ (so they can swing freely) and hammer flat.

Divide each lid into 16 equal sections. Notch each tip except the one from which the star will hang; cut that to a point and strike an awl hole in it.

When all 16 stars are cut, spray-paint them with Krylon Flat White Enamel. Set aside.

SUN PENDANTS

Cut down the 16 frozen juice cans to within 1″ of the bottom.

Divide the cut-down sides into 8 equal sections. I must confess I did this by eye rather than using the folding-paper technique, because I felt the chandelier was just a gay thing which did not warrant an undue amount of slavish accuracy and care. I also cut the curves in each section freehand.

Hammer the sections out to the side; strike the awl holes; and spray-paint with Krylon Bright Gold Enamel.

Form links by winding florist wire around a dowel or large knitting needle and cutting through the coil. Link the stars you have made to the sun pendants and the pendants to the chandelier.

WIRE THE CHANDELIER

Decide on the height at which you wish your chandelier to hang. It would be lower, for instance, if used over the dining table than it would be if hung in the front hall. Figure out how long the chain must be and count the number of links to make sure the three lengths are identical.

Wire the chain to the chandelier and suspend from a hook in the ceiling. As you can see, I wired on a wreath and ornate Imperial Star as well. After all, Christmas comes but once a year!

SUGGESTION

Make a lantern out of your chandelier by piercing the center circle (the bottom of the can) and wiring a pierced cone on it, concealing a light bulb.

REFLECTIONS AND INSPIRATIONS

No other group of people has realized the artistic potential of tin more fully than the early Spanish Americans. Fervent Catholics, they seized upon this shining raw material as a substitute for silver in fashioning their altarpieces, crosses, and candlesticks. In their homes, also, hung the image of the Virgin and the owner's patron saint, often painted on tin and framed in it. The very simplest of these frames, unsoldered, was a sheet of tin with tabs cut around the edge, every other tab bent over the front to hold the painting in place. More elaborate frames involved the soldering on of punched panels, crimped fans or foliage at the corners, and radiant lunettes overhead.

The Spanish Americans also enhanced the sparkle of tin with glass. The more skilled of them painted flowers and birds on the reverse side of the glass to glow against the tin backing. (See drawing of Spanish-American cross in diagram on page 124.) Others less skilled obtained graceful effects by coating the glass with paint and raking it with a comb when tacky, in patterns of waves and scallops, through which the tin shone. Here, then, is a collection of designs for your reflection and inspiration.

MEXICAN SUN MIRROR

One of the most popular mirror forms ever designed, and it couldn't be simpler to make.

TOOLS
Snips (even scissors)
Paring knife
Compass

MATERIALS
2 unribbed 2-pound gold cans
Weldit Cement or Epoxy 220
Shelf paper, ruler, pencil
Rubber cement
Circular mirror
Pull-chain beading *(optional)*

If you are lucky enough to have a fish man, he will probably get scallops in these gold cans.

Divide the sides of one can into equal sections (see

Dividing Sides of Cans Into Equal Sections, page 15), *ruling off the rays* on the folded paper before applying it with rubber cement to the can.

Cut the rays (Fig. 1) with snips or kitchen shears; remove the paper and the glue; and, from the inside, hammer the rays out to form a halo.

Make two holes in the bottom of the can (Fig. 2) and wire for hanging.

On the second can, mark a line with the compass (see Cutting Down Cans, page 14) 1½" below the top rim. Starting near the seam, cut along that line with the paring knife (see Removing Rims, page 14), thus dividing the can into two sections. The smaller section will form the collar and the larger will make a second set of rays.

Divide the sides of both these pieces into equal sections as you did with the first can, and hammer the rays out similarly.

Glue the second set of rays to the first. Glue on the mirror cut to fit inside the rim. Glue on the collar over the mirror. Glue on bold pull-chain beading if desired.

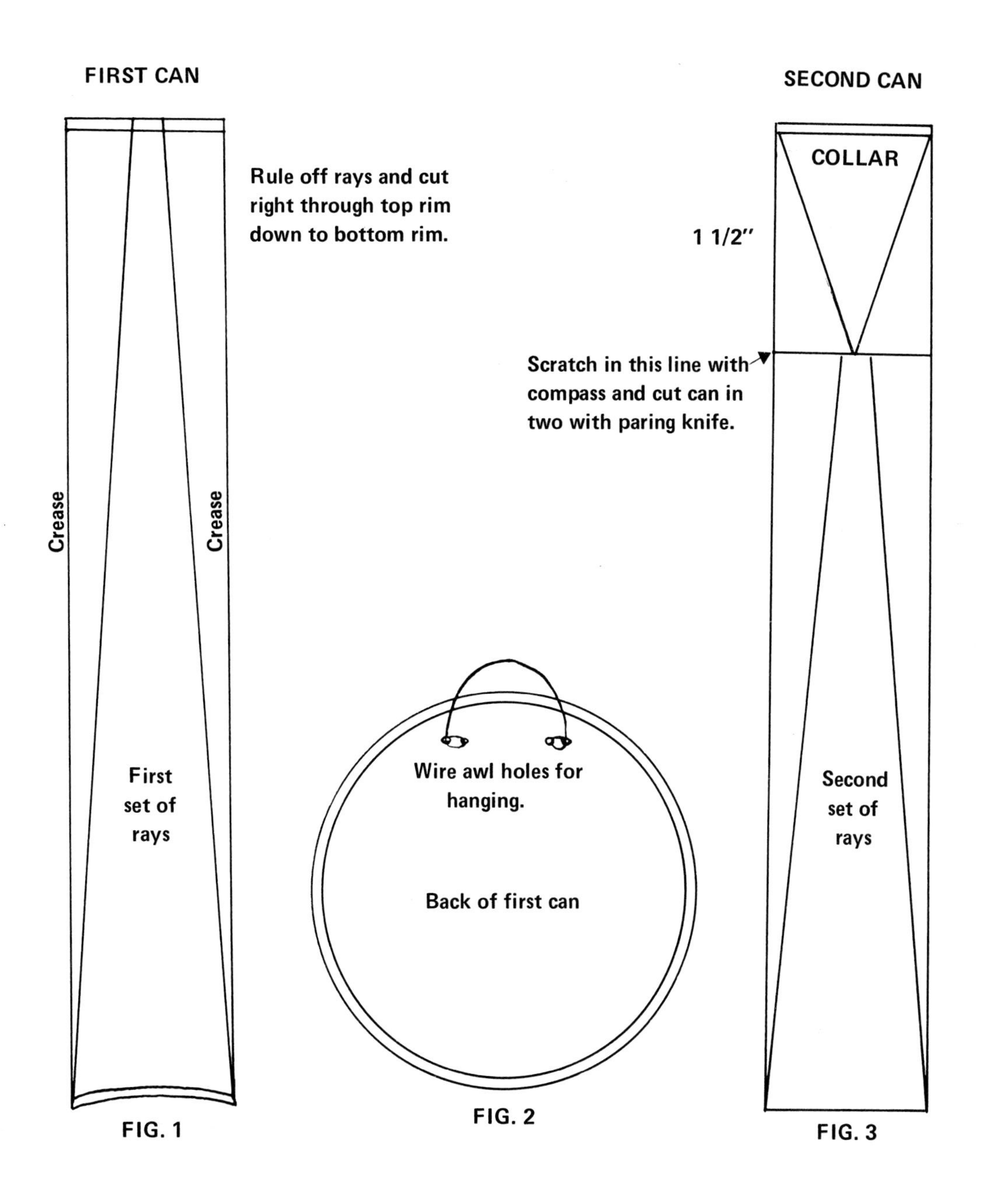

FIG. 1 FIG. 2 FIG. 3

THE FRAME

If you do want the Frosty Star to brighten the mirror at Christmastime, remove the bottom of the can.

If not, leave it in place and, in any case, set about fringing the sides of the can (see Fringing Tin, page 15). Should your can be more than 4″ deep, cut it down (see Cutting Down Cans, page 14) unless you like the idea of framing the mirror with spirals like those on the Morning Star on page 184.

As you fringe, hold the third finger of your helping hand *in the way of the curls* as they curve down, so they won't corkscrew tightly.

That's all there is to it. Glue a wired tab on the back for hanging (or use the Mexican Sun Mirror method, preceding); glue the mirror in place; and your *chef-d'oeuvre* is complete!

THE STAR

Having removed the bottom lid of the can and fringed the sides, glue the top lid of the can, concave side up, to the underside of the frame with Epoxy 220.

Cutting the Frosty Star involves the same sort of idiot's delight type of technique as the frame, i.e., cutting curls.

Simply divide the lid into 16 sections and make narrow cuts down both sides of each section

FROSTY STAR MIRROR

Five o'clock of a Friday morning finds my fishmonger at the Fulton Fish Market, deep in the heart of New York City, selecting succulent scallops and sweet flounder fillets to bring home to Fairfield, Connecticut. All unwitting, he transports them in fabulous tin cans, never realizing the ravishing uses they might be put to, and consequently throws them out or gives them to me gladly. I suspect all fishmongers do the same. Why don't you ask yours?

Complicated though it may look, this mirror is simplicity itself. Try it and see. You will need the whole can, including the lid, unless you would rather have just the mirror without the star.

TOOLS
Coarsely-serrated snips
Hammer

MATERIALS
Shallow commercial fish can, 4″ x 12″
Mirror cut to fit
Weldit Cement or Epoxy 220
Felt-tip pen *(optional)*

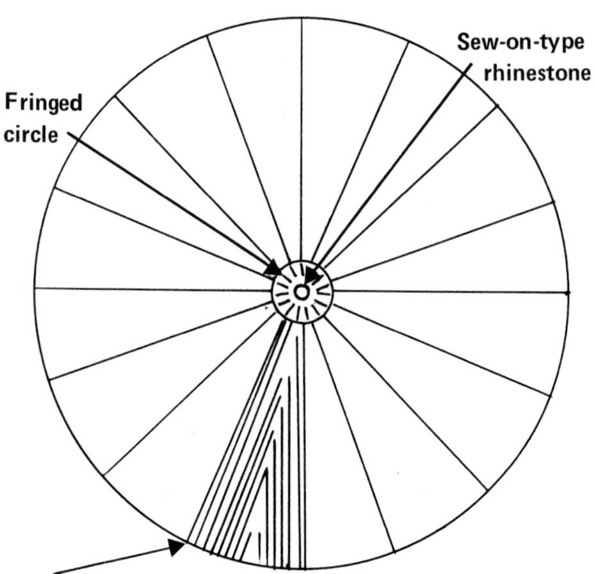

With coarsely-serrated snips, divide the lid into sixteen sections. Cut to imaginary line on both sides of each section.

to an imaginary line (see diagram). Nothing could be easier, but I will give you a few tips.

In dividing a lid this large you may want to find its center with the compass, and rule off the lines, or at least the first four, with a felt-tip pen. (Be sure to remove the lines with alcohol before cutting the curls.)

It is also best to glue the center ornament on the star *before* cutting the sections. Since this star is so much larger than average, I use two fringed circles in the center instead of one, the larger from a small all-silver tomato paste can, and the smaller from a circle the size of a quarter.

In cutting the strips down each side of the imaginary line, let them curl as much as they wish. Don't inhibit them this time, as you did with the frame. In fact, encourage each curl to spiral off to the right by pushing it gently into place if it seems at all reluctant. Work around the star to the left; hold each section level as you cut it; and keep your snips perfectly straight. Then all should curl down, up, and away!

Glue a 4″ wired tab for hanging on the back of the lid with Epoxy 220, and let dry overnight.

The next morning, glue the mirror in place. Wire the star to the fringe at center top and bottom.

MEXICALI ROSE MIRROR

This looks quite romantic, almost more Victorian than Mexican, probably because of the naturalistic treatment of the leaves, but the roses are actual copies of the stylized Mexican ones. The frame is nothing more than the whole end of a 5-gallon soybean oil can, with ornamented rims soldered on.

TOOLS

Snips
Ball-pein hammer and board
File
Long-nosed pliers
Screwdriver
Chisel *(optional)*
Nailset
Awl *(optional)*
Propane torch *(optional)*
Asbestos mat *(optional)*

MATERIALS

Square 5-gallon can, 9″ wide and 13½″ tall
4 lids from small frozen juice cans
Rims of large ham or fish can
Architect's Linen, scissors, pencil
Rubber cement
0000 steel wool
Epoxy 220
Solder, Dunton's Tinner's Fluid *(optional)*
Florist wire
Mirror

THE FRAME

Remove the whole end of a square 5-gallon can (see Removing Rims, page 14) and file it smooth. Burnish well with steel wool.

Order mirror from your local glass-cutter to fit center circle.

Cut four roses from lids of small frozen juice cans.

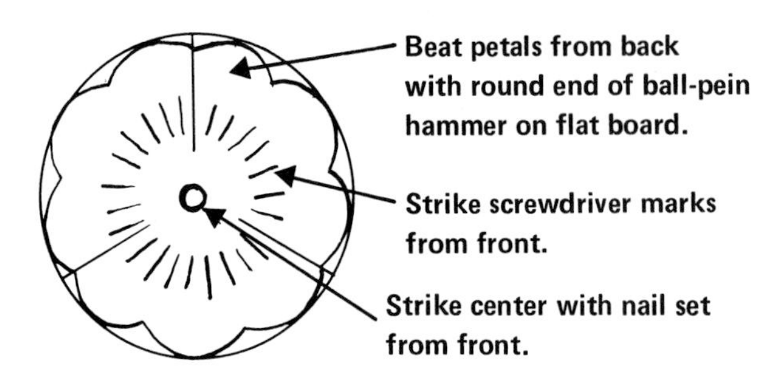

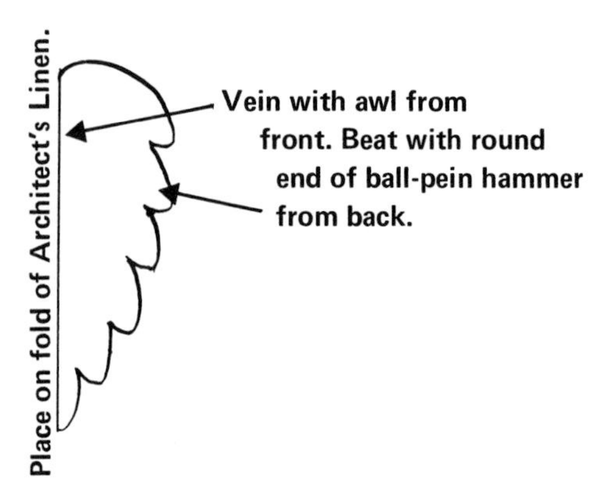

THE ROSES

Cut out four roses and eight leaves, applying the Architect's Linen pattern to the side of the can with rubber cement. Hammer them flat.

Beat the rose petals with the round end of the ball-pein hammer from the back on a flat board.

Strike the screwdriver marks at the base of each petal from the *front.*

Strike the center mark with the nailset from the front.

Inscribe a central vein in the leaves from the front with awl or chisel.

Turn the leaves over and hammer on either side of the vein with the round end of the ball-pein hammer on a flat board.

Glue or solder them in place on the corners.

Ornament the rims from as large a circular ham or fish can as you can find (see Ornamenting Rims, page 15). Bend them to fit the circles, using long-nosed pliers if necessary. The larger rim may be either glued or soldered in place, but the smaller must be glued on after cementing the mirror in the center.

Glue or solder on a wired tab for hanging in the back.

In soldering, the mirror acquires an antiqued pewter color, and you may wish to leave it that way. The Mexicans, of course, would use vivid transparent paint, which you can approximate with Dri-Mark permanent pens. However, the most permanent transparent paints are the Talens Glass Paints or artists' oil paints diluted 50 percent with spar varnish. My mirror, though, was painted impressionistically in opaque oils by Ann Laredo.

EVENING STAR MIRROR

Cans in general, and stars in particular, lend themselves ideally for use as mirrors or frames. The classic curve of the feathers in the Evening Star, enhanced by the satin sheen of silver graced by gold, makes this design a constant pleasure to live with.

TOOLS

Serrated snips
Hammer and board

MATERIALS

V-8 Juice can
Rim from another V-8 Juice can
Weldit Cement
Florist wire
Mirror
Krylon Crystal Clear Acrylic Spray

Leaving the bottom lid of the V-8 Juice can in place, follow the directions for the Evening Star on page 182.

Replace the Frosty Star in the center with a mirror cut to fit.

Do not bother to spray-paint the ornamented rim gold, but do spray the star itself with Krylon Crystal Clear Acrylic that it may never, ever, lack luster.

For other mirror ideas, see the following drawings of two Mexican frames: the Santa Fe Looking Glass and the Tarascan Reredos.

TWO MEXICAN FRAMES

SANTA FE LOOKING GLASS

Single dots from back with nail set. Stars from back with chisel. Outline of hearts and leaves from top with awl, veining from back with chisel.

TARASCAN REREDOS

Frame upon frame, all tin. Crosshatch on outer frame first struck from back with chisel, then raised with ball-pein hammer.

A MEXICAN RETABLO

The Mayans and Aztecs notwithstanding, Mexico is a country of cathedrals—the Spanish saw to that. Whatever else it may lack, even the smallest village has its twin-towered *iglesia,* giving on the square. And this is where the action is—in the church and *jardín;* this is the heart of Mexican life. All of which you realize so much the more when you enter the chapels rimming the cathedral. There the humble and devout bring offerings of thanks to their saints. The walls are lined with *retablos* that tell the grateful stories of the miracles the saints have wrought in their lives. Usually the saint has saved a life, and the painting will depict a bare room with a still, pale figure stretched on a pallet under the radiant image of the saint, to whom the members of the family are kneeling in supplication. Another may show the owner of five lost bulls, kneeling outside his hut, praying to the vision of his saint who, miraculously, restores them. Usually, under the scene depicted, the details of the story are penned in black ink, dated, and signed. Almost always the *retablos* are painted on a piece of tin, frequently extracted from the side of a can. So also with their frames, if they have any. Crudely executed like the paintings, they are genuine expressions of reverence and devotion.

Working in that same spirit, I have made a frame for my saint from a battered fish can and beer bottle caps. (See color photograph on page 34.) Rather than giving you step-by-step procedures for this, since your needs would be different, I'll give you the basic approach that I used.

Even though the Mexicans do not, I mounted my *retablo* on a piece of ¼" plywood, using small brass nails which I bent over on the back. There is a long European tradition for this method, and it eliminates soldering. I probably should have nailed the beer bottle caps on too, but they were such flawless seals, I couldn't bear to puncture them, and glued them on each of the four corners with Weldit Cement.

The lunette at the top is a typical Mexican motif, derived from the shell form, called *concha,* which the Spanish favored. It is invariably used in the framing of *retablos* and for many other things besides, such as chair backs, niches, portals, etcetera. Mine is made from half of the bottom lid of a large commercial fish can. As you can see, it capitalizes on the design contributed by the concentric circles, and augments it with rays struck with hammer and cold chisel.

On the top band of the frame proper there is a beading which I particularly like. This just happened when I was forced to remove the bottom lid with an old-fashioned can opener. As I was working, I saw to my dismay that the opener was cleaving the rim. Then, to my amazement, I noticed that it was producing a patterned edge that was actually decorative. In such witless ways are useful discoveries made!

The basic frame is made of four strips of tin taken from the sides of the can. The inner side of each strip is folded over as if making a scroll. The other side is wrapped around the back of the board and hammered against it. The strips are nailed on through awl holes made before mounting them on the wood backing. The loop for hanging is made from the rim of a can, hooked into two screw-eyes. *¿Bellísimo, no?*

SPANISH-AMERICAN CROSS

Reverse painting on glass panels, superimposed on tin frame. Could be a pendant, or wall sculpture, depending on size.

ICONS

Early in the Christian Era, icons were somber, naïve images, wholly unadorned, but as the Church grew in wealth and power, all the outward manifestations of the faith became increasingly elaborate, icons not excepted. Gold and precious jewels were superimposed upon the frescoed figures until nothing but their faces showed. Most of us, I think, rather like the contrast between the primitive, stylized features and the later Byzantine embellishments, and, if you study an icon closely, you will be intrigued to discover what simple devices have been employed to achieve that richness. Many fine borders consist of nothing more than a tiny beading made with a tap of the awl from the back. All that is required in making an icon is care and persistence. Striking all those little blows is tedious, but after all, you don't have to do it in one sitting. In spite of its elaborations, the frame for the colorful advertisement you see here only took three hours to make.

TOOLS

Snips
Awl
Hammer and board
Screwdrivers of different sizes
Chisels
Nails
Oval die
Nailset

MATERIALS

Sides of cans, unrimmed and lightweight
Small round-headed brass nails
Upholsterer's tacks
Steel wool
Weldit Cement or Epoxy 220
Ruler and Posey Clay
Tracing paper, carbon paper, lithopone, ballpoint pen
Talens Transparent Glass Paints
Cabochon glass stones
Stained board

ICONS

BORDER DESIGNS

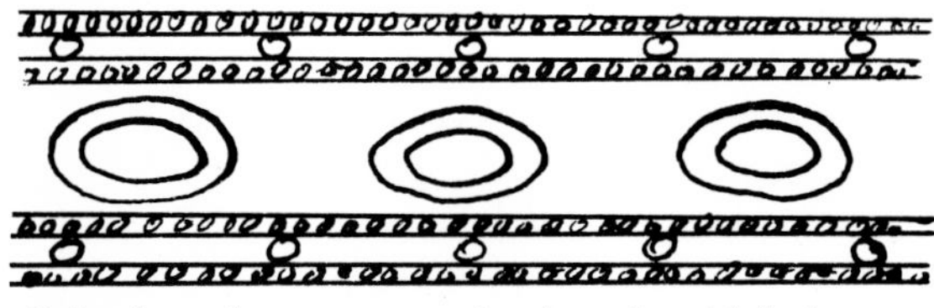

Cabuchon glass stones under tin strip with holes; brass nails; awl beading from back.

Sawed-off nail or screwdriver from back.

Oval die from front; awl from back.

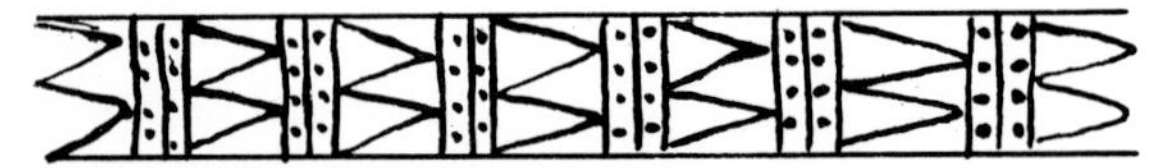

Bands made with chisel from front and awl from back. W made with chisel from back.

Bands made with chisel from back. Stars made with screwdriver from back, sawed-off nail from front.

Nail set from back; oval die from front.

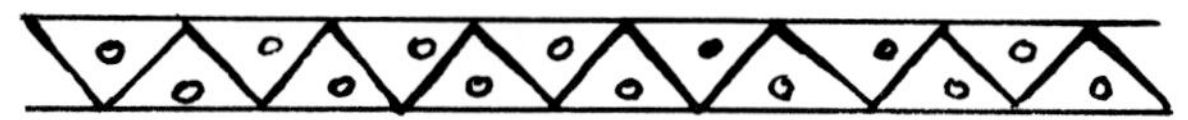

Nail set from back; chisel from front.

Oval die from back; nail set from front.

Cut trefoil pattern; apply with rubber cement; draw stem; stipple with awl from back.

Chisel from back; screwdriver from front.

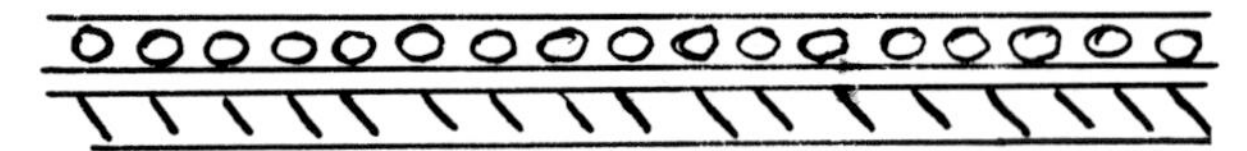

Nail set from back; screwdriver from front.

Draw serpentine line with awl first. Mark off pattern areas with felt-tip pen and draw design. Strike all leaves and strawberries from back with oval die and awl. Strike petals from front with oval die; strike flower center from back with nail set.

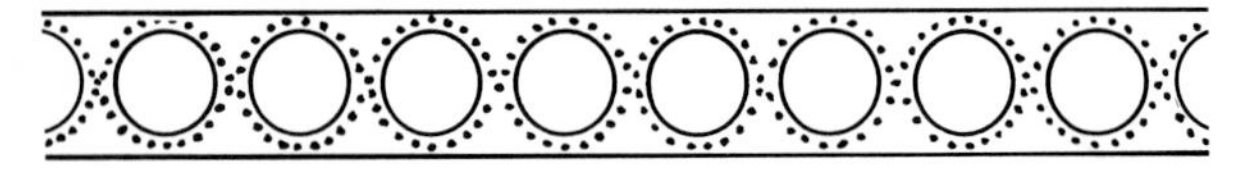

Toothpaste cap from top; light tap of awl from back.

As with the Mexican *Retablo,* I shall give you a general procedure to follow, since sizes and designs would be different in each case.

First, cut the basic frame to fit the image, and burnish it bright with *0000* steel wool.

Make four awl holes in the corners and tack the frame to a clean hardwood hammering board, lightly enough so that you can remove it easily, but securely enough so it won't pop loose.

With awl and ruler, score the border lines. Two dabs of Posey Clay on the bottom of the ruler will help to keep it in place.

If nailheads are to be an integral part of your design, make the necessary measurements and strike the awl holes.

Lift the frame off the board, turn it over, and tack it on again. Strike all the awl holes again from the back and hammer them flat.

If you plan to use a repeat pattern on the inner and outermost edges of the frame, such as a braid or bead, strike that now from the back. It will then appear raised when viewed from the front.

Any interior design should be worked out on paper ruled to size, unless you are going to use one of those provided here, in which case trace it off, if necessary. Most of them can be executed freehand, but there are some which should be measured and marked off with a felt-tip pen. Transfer the design to the tin either with carbon paper and ballpoint pen, or with lithopone and sharpened pencil. Transfer to the *back* of the frame only those parts of the design to be *raised.* Transfer the remainder of the design to the front of the frame.

Mount your icon on a board cut to fit, sanded well, and darkened with stain, shoe polish, burnt umber, or whatever. Wire it for hanging.

ACCENTS AND THINGS

Don't you find yourself, every now and again, complimenting a friend by saying that you love her little touches? And even as you call them "little," you are declaring their importance. For it's those little touches that distinguish her as an individual, put her personal stamp on the house, make it uniquely her own. Never essential items, they make an essential difference.

In this section you will find designs for key places in your home—the front door, the windows, hall, or fireplace. They are not essential, they are strictly superfluous—but they make an essential difference.

FEDERAL FANLIGHT

No doubt, opportunities to use this design will be rare, but just as I had need of it, so may a few of you. To most, however, I offer it as a suggestion—and so with the Mexican Fanlight too. For instance, this Federal Fanlight could be equally suitably used as a bas relief on a valance. It would make an effective ornament on a planter or window box. Differently combined, the rosette and bellflower festoon could be adapted to bathroom or kitchen cabinets in lieu of provincial moldings. Then, there is always the possibility that you might one day find yourself remodeling a house and in need of an architectural feature to indicate the period you have in mind. I was delighted to discover that inserting this fanlight over my front door immediately made my twentieth-century house seem as eighteenth-century as the furniture inside it.

Here, then, for possible future use, is my adaptation of a fanlight taken from the home of Mr. and Mrs. Robert Carter in Old Lyme, Connecticut. The original design included double-oak leaves, running vertically in the open spaces between the two large bellflowers. Feeling that they were too much for my small window, I used the design as a hinge on the Heart Keepsake, page 154. You'll find the pattern for it there, in case you want to use it in its proper place. I made two identical fanlights, one the mirror image of the other, and glued them to the window with Weldit Cement. You could substitute Epoxy 220 for the solder, but not for the Weldit Cement; you might never be able to get it off should you change your mind.

The pattern for the Mexican Fanlight is also given here.

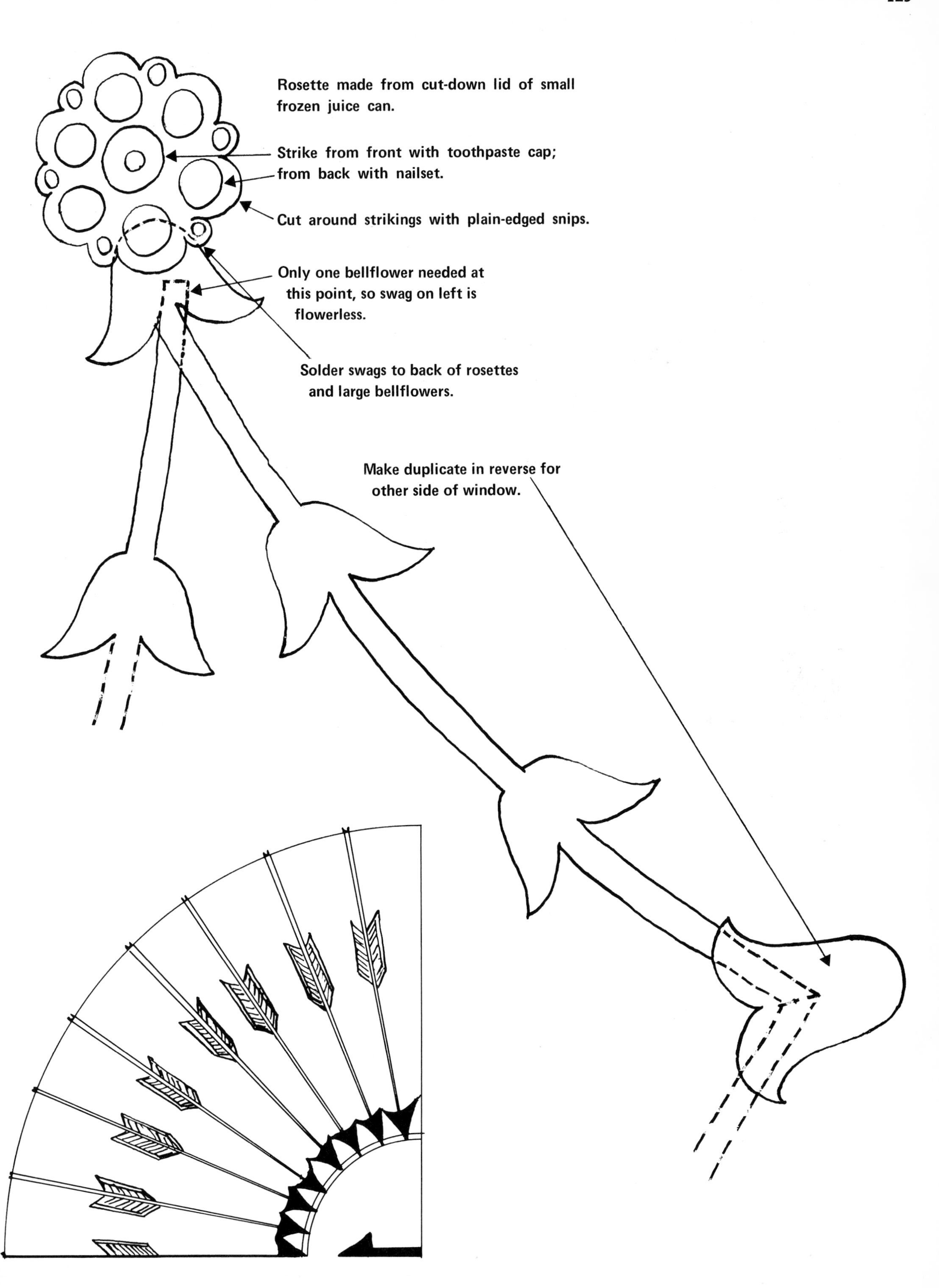
Rosette made from cut-down lid of small frozen juice can.
Strike from front with toothpaste cap; from back with nailset.
Cut around strikings with plain-edged snips.
Only one bellflower needed at this point, so swag on left is flowerless.
Solder swags to back of rosettes and large bellflowers.
Make duplicate in reverse for other side of window.

CANAPEAR CURTAIN

Handsome as a room divider, especially in a contemporary setting, but harmonious even with the traditional, this curtain is illustrated in color on page 102.

TOOLS

Plain-edged snips
Hammer and board
Awl
Long-nosed pliers

MATERIALS

82 identical, unribbed soup cans with rich gold linings (at least)
Florist, or Nu-Gold, wire
Decorative brass curtain rod, and brass cup hooks to hold it
Architect's Linen, scissors, pencil
Rubber cement
Weldit Cement

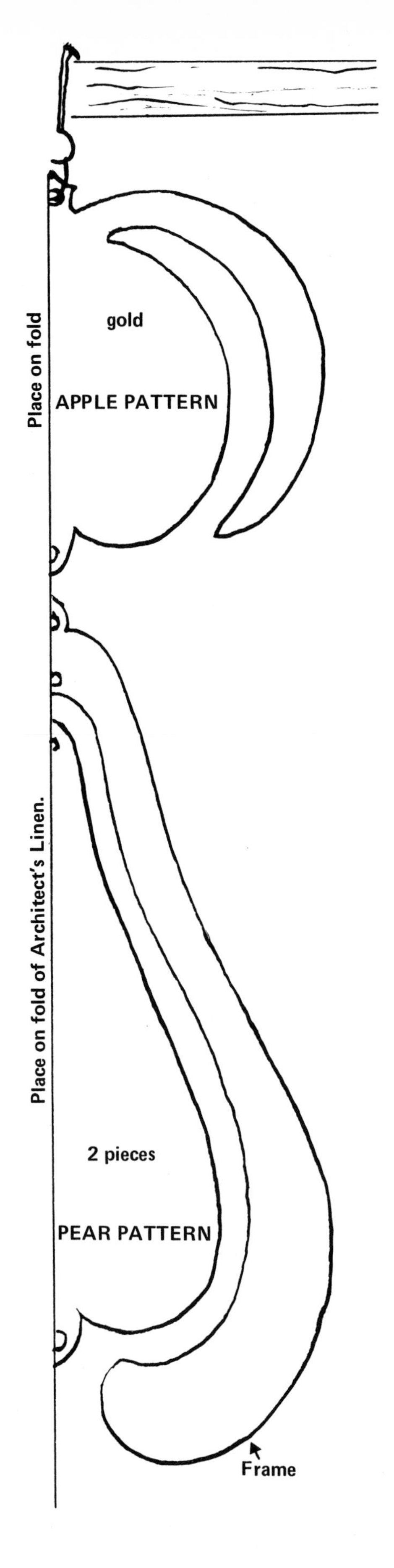

A WORD ABOUT THE CANS

As a confirmed Campbell's Soup-can fancier, let me point out the advantages. The size is right (the pear design fits perfectly, with just the necessary ⅛" to spare). The weight is right, a lovely, limber piece of tin. The color is right, the outside being a frosted silver, and the inside, the richest gold available. Shop for Beef Bouillon, Black Bean, Consommé, Green Pea, Golden Mushroom, Hot Dog Bean, Minestrone, Onion, and Pepper Pot. There are undoubtedly others that I haven't yet discovered. Unfortunately, none of the cream soups, which are so handy as a base for sauces and gravies, are aggressive enough to require lacquering, and the beauty of the curtain depends in good measure upon the contrasting of silver and gold. But the list above will give you *and* your friends (and you'll soon know who *they* are!) quite a variety of taste sensations to choose from. It will only take all of you a month or so (of persistent shlurping) to acquire enough cans for this curtain.

THE PEARS

As you can see from the diagram, the pears are composed of two separate pieces: the pear itself and a frame.

Architect's Linen is a must for this project, where the pattern is used so many times. In fact, you will need two pear patterns before you are through, so you may as well trace them both off at the start. Do it accurately; the proportions of this design have been worked out quite carefully.

After opening up the can and hammering the edges, coat the gold side with rubber cement and center the pear on it.

Then center the frame directly over the pear (as indicated in the pattern diagram), taking care to keep it equidistant all the way around. This will ensure a true frame.

Cut out with plain-edged snips, slowly and fastidiously. (It's amazing what a difference 1/32″ can make.) To assist you, I have made a separate pear-cutting diagram, outlining the route that seems best to me. Those of you who are left-handed would reverse it.

Remove the pattern and the glue, and hammer flat. Improve the outline, and hammer flat again.

Strike the awl holes, making sure that they are *all directly under each other;* otherwise the frame will kick off to the side.

Hammer both the pear and the frame from both sides until they lie *perfectly flat* on the board.

Make rather large links by coiling florist or Nu-Gold wire around a 1/4″ dowel or knitting needle, and cutting through the coil.

Assemble the individual pear units, linking a golden pear to a silver frame. Apply a drop of Weldit Cement with a toothpick to the joining of each link.

Pear-cutting

Cut into corner here, then withdraw and bypass the corner, coming back to it later.

y

Start here.

Cut curve to here. Then cross over into corner in base of pear; withdraw and bypass, etc. Cut out pear, then finish cutting frame along double arrows, starting at y.

THE APPLE

Trace the pattern on a fold of Architect's Linen and cut four apples. Remove the pattern and the glue; hammer flat. Improve the outline and hammer flat again.

Strike the awl holes in the stems from both sides and hammer flat.

ASSEMBLE THE CURTAIN

Suspend the curtain rod, strung with five curtain rings, from two brass cup hooks.

Starting at one side, link 16 pear units, one under the other, alternating gold and silver frames. Begin the next row with an apple. Add 15 pears, alternating gold and silver frames with those in the first row. Finish the second row with an apple.

Repeat this process three more times, taking special care to keep the pears *in every other row level* with each other, so that the diaper pattern is perfect and pleasing.

ASSEMBLY

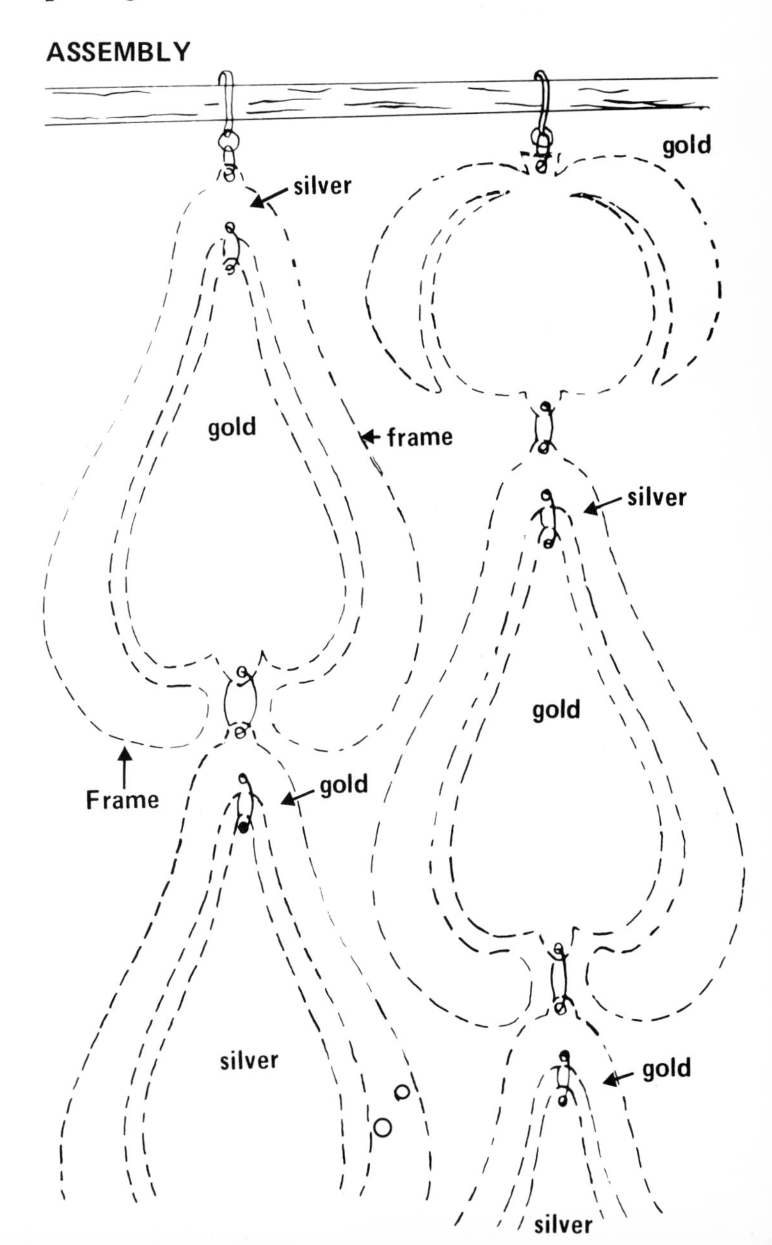

IMPERIAL FIREPLACE FAN

Easy and elegant—a smashing gift. All you need are coarsely-serrated snips and a 4-gallon can from your local delicatessen. Ask if you might have the potato salad can, or the one fresh blueberries come in; they are usually all-gold, and go well with brass fireplace fittings.

TOOLS

Coarsely-serrated snips
Hammer and board
Awl *(optional)*
Propane torch *(optional)*

MATERIALS

4-gallon can, preferably round, measuring 10″ wide and 13″ high
Weldit Cement or solder, Dunton's Tinner's Fluid
String, felt-tip pen, turpentine
5 upholsterer's tacks
Ruler, grease pencil
Steel wool

CUT OUT THE FAN

Remove the bottom lid of the can. (Save it. It will be the center ornament of the fan.)

Open up the can (see Opening Out Cans and Cutting Curved Ribs, page 14); step on it; and hammer it flat. Burnish it bright with steel wool, unless it is gold, in which case wash it with a soft soapy cloth.

Find the center of the long side of the flattened sheet and mark it with an awl hole large enough for the string to pass through. Inscribe an arc as high as the can with a felt-tip pen tied to the string. Inscribe a second arc two inches smaller than the first (Fig. 1).

Cut out the fan (along the outer arc) with coarsely-serrated snips. With ruler and felt-tip pen or grease pencil, divide the half-moon into sixteen equal sections (see Dividing Lids Into Equal Sections, page 15) by first dividing it in half, then in quarters, then eighths, then sixteenths (Fig. 2). You can do this pretty well just by eye even though the circle is large. If you have a large protractor you may want to use it, but mine was so small that the rulings didn't come out any more accurately than if I'd done it freehand, and I had to even up the sections with a ruler. You'll find the sections are about 2″ wide on the inner half-moon.

Cut skinny wedges out of the perimeter according to the diagram to make true square sections for fringing. Draw scallops in the sections, freehand, with felt-tip pen. Cut fringe *to* the scallops, holding the third finger of your helping hand in the way as it curls down to keep it more nearly straight.

When you have all the sections fringed, follow the how-to procedure for the Imperial Necklace, page 174. There I have tried, with drawings and instructions, to clarify the simple process of rounding the fringe backward, pulling each strand out to the right (or the left), and putting little hooks on the tips with the round-nosed pliers. If you will study both the photographs and the drawings, you'll soon catch on. It is a very simple and most effective device.

A few suggestions about curling the fringe on the fan: work *every other* section from the front; the alternate sections from the back. Do all similar operations at once, that is, all curving of fringe, all pulling to the side, all hooking, etcetera. You will achieve a sort of rhythm that way, and it will seem to go along faster.

When your lace-edge is finished, you may want to emphasize the sections further by ruling in the rays with the awl. This will provide a certain architectural quality to the fan that is pleasing and make it seem more substantial.

Fold the bottom lid of the can in half, by bending it over a counter edge. Center the fan in the folded lid and glue or solder in position. Add upholsterer's tacks for ornamentation if you like, striking the necessary awl holes in the fan, clipping off their sharp points, and gluing (or soldering) them in place.

THE FEET

You don't actually need feet: a paper fan would not have any. You would simply place the fan on the andirons behind the finials. But I rather liked the idea of having the fan free-standing, and made curved feet from two 18" x 1½" strips of tin taken from the flattened side of a 12-pound ham can. If you don't have such a can, you can make triangular feet from the tin you have left over from the fan (Fig. 1). The strips should be about 13½" x 1½". Fold the edges of the two strips as you would if making a scroll.

Round the ends as indicated in Fig. 1. Bend the strips as shown in Fig. 3. Strike an awl hole where you plan to secure the feet to the fan with an upholsterer's tack. Hold the feet in position on the fan and mark the spots for the awl holes. Strike them, and glue (or solder) all together.

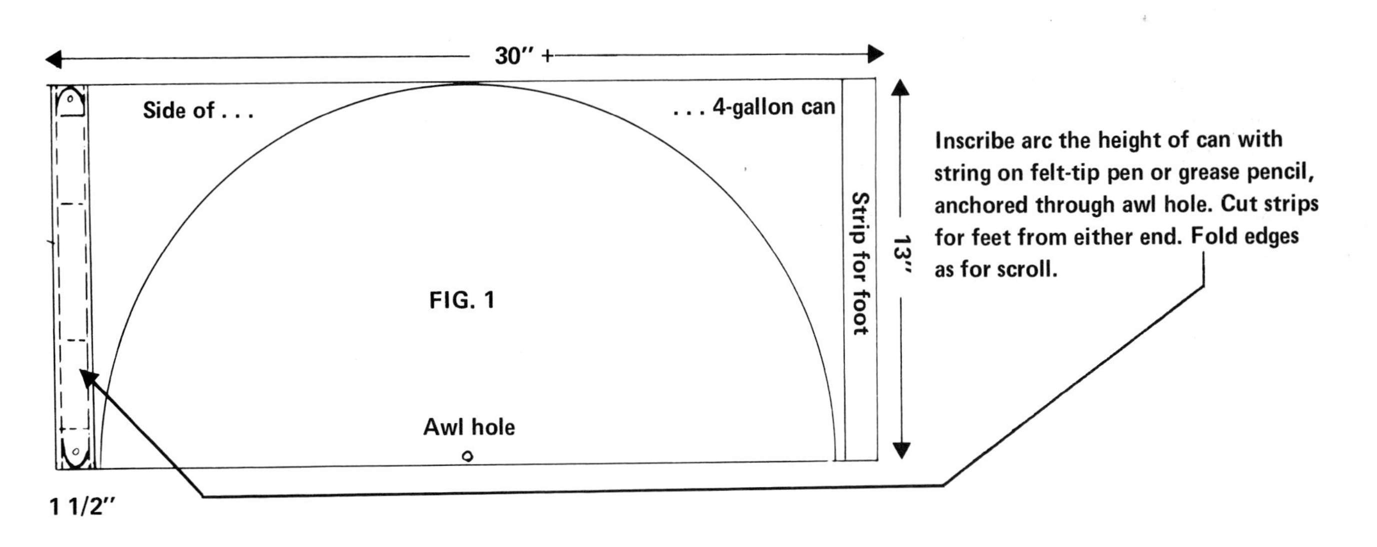

FIG. 1

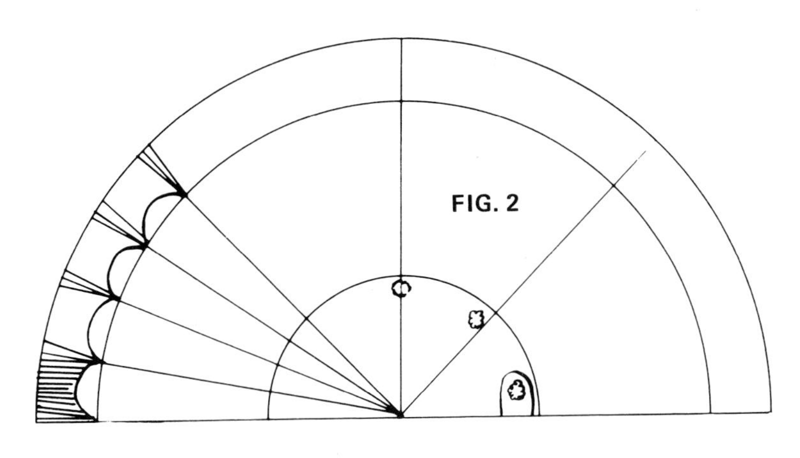

FIG. 2

Inscribe second arc 2" inside first. Divide fan into 16 equal sections by eye. Describe small arcs in each section with felt-tip pen. Remove wedges as shown and fringe each section down to the arcs. Follow directions for Imperial Necklace. Fold can bottom in half and glue (or solder) in center. Secure feet with glue (solder) and upholsterer's tacks.

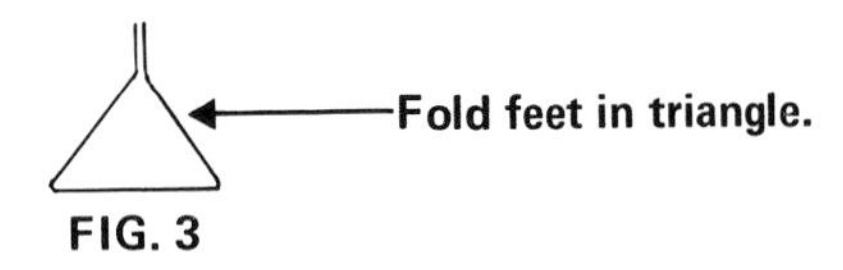

FIG. 3

LAMBREQUIN

Time was when the term lambrequin referred to the veil that knights wore over their helmets as protection against the elements, or the hood placed over the head of a falcon, but now it refers to a stylized architectural drapery, framing a window or bed. It has the virtue of doing the job economically and aesthetically; it takes up no room; and doesn't have to be sent to the cleaners.

I had already been planning to make tin curtains just for fun, when one day, after cutting out the Imperial Fireplace Fan (preceding), it dawned on me that the two end-pieces of the 4-gallon can would fit perfectly into the corners of my desk window. So I tacked them in place, hung my favorite five-pointed star in the center, and there I was with a lambrequin. Since then, I've come up with a number of variations which may inspire you to make your own. Ann Laredo put her inimitable touch on the one you see here, this time with gay Dri-Mark pens.

DESIGNS

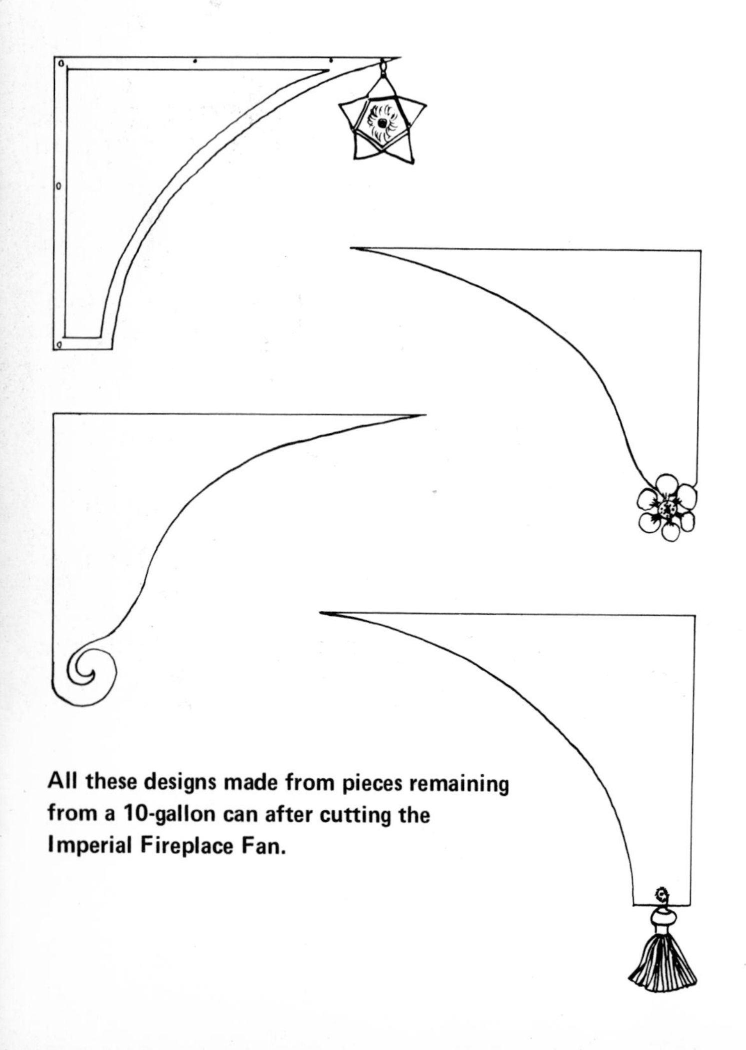

All these designs made from pieces remaining from a 10-gallon can after cutting the Imperial Fireplace Fan.

These three designs made from the entire side of a 10-gallon can. May also be pierced in manner of 18th-century Venetian blind valances, or punched as in Early American pie-safes and lanterns.

Tin is suited to perfection for Christmas decorations, as shown by the Festive Swag over the mantlepiece in Lucy Sargent's home. The garland is ornamented by the Cheerful Cherub in the center, crossed ribbands garnished by little rosettes and two rose tiebacks at the top corners. The two tassels at each end are mounted on champagne corks.

Created from lowly frozen-juice cans as if by magic, the beautiful Star of Bethlehem has a particularly fitting place in the humble Nativity Scene and the miracle of Christmas.

PINE-ROSE TIEBACK

In an effort to reproduce in tin the cross-sections of the yellow pine cone (See Catalyst—An Assemblage, page 64), I came up with this very useful rosette. It's good in all sizes.

TOOLS

Plain-edged snips
Ball-pein hammer and board
Chisel
Awl
Nailset

MATERIALS

Lids from coffee cans
Mirror rosette or screw
Grease pencil or felt-tip pen
Steel wool

Decide upon the size of the lid which will best suit the specific place you have in mind; remove any price mark, and burnish it with the steel wool.

Do *not* hammer it, but go directly to work dividing it into eight equal sections (see Dividing Lids Into Equal Sections, page 15).

Trim the sections as indicated, drawing the design on with grease pencil or felt-tip pen, if necessary.

Strike the awl hole in the center from the top, then the back, and hammer flat from the back.

With hammer and chisel, strike the parallel lines at the base of the petals.

Beat the petals and the center from the *back* with the round end of the ball-pein hammer on a flat board.

Stipple the center with a nail set from the back.

Turn over and beat the chisel marks with the round end of the ball-pein hammer from the front.

Paint the rose, if you like. In the Festive Swag, page 185, they are mustard and silver; they would be very effective in antique gold and white, using the original markings on the cone scales. Screw to the wall with the mirror rosette.

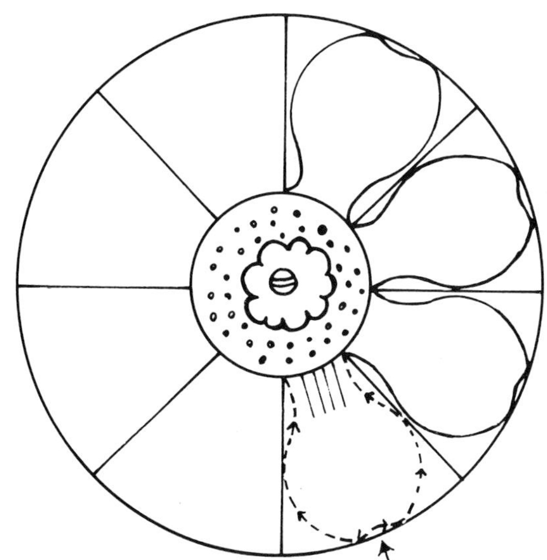

Cut in both directions from this point.

Divide lid into eight equal sections. Draw on design with grease pencil and cut each section as indicated by the arrows.

Make the awl hole in center; strike chisel marks at base of petals from front.

Beat center area and petals from back with round end of ball-pein hammer on a flat board.

Stipple center from back with nail set. Turn piece over and beat chisel marks from front with round end of ball-pein hammer.

Screw on wall with rosette.

HANDSOME HOOKS

All sorts of stylized flora and fauna make handsome hooks, especially the butterfly with his wiry feelers. This is another nifty gift.

TOOLS
Plain-edged snips
Round-nosed pliers
Long-nosed pliers
Ball-pein hammer and board
Propane torch
Asbestos mat

MATERIALS
Assorted lids and sides of cans
Enamel paints
Solder, Dunton's Tinner's Fluid
16-gauge wire or brass welding rods
Architect's Linen, scissors, pencil
Rubber cement
Mirror rosettes or screws

Unless you intend to paint the hook, choose a can with a rich gold lining and a heavy weight of tin, like V-8 Juice. You'll have to hammer the ribs flat, of course, before applying the patterns, because a symmetrical effect counts for a lot in these designs. (Glue the pattern together before cutting it out; see Cutting Out Patterned Designs, page 16.)

You may want to achieve a sense of third dimension and yet have the ornament lie flush against the wall, so beat it from the back with the round end of the ball-pein hammer on a flat board.

A 10″ length of 16-gauge wire (or brass welding rod of equal gauge), folded in half, squeezed tight with the long-nosed pliers, and bent around a broomstick is about right for most hooks. You will want the wire to travel up the back of the design as far as possible in order to strengthen the hook, but on the other hand, the screw has to be worked into the design along the way, so think about that before soldering the wire on. Mirror rosettes could be used to fasten the hooks to the wall, but would be too deep for doors.

FLEUR-DE-LIS PATTERN

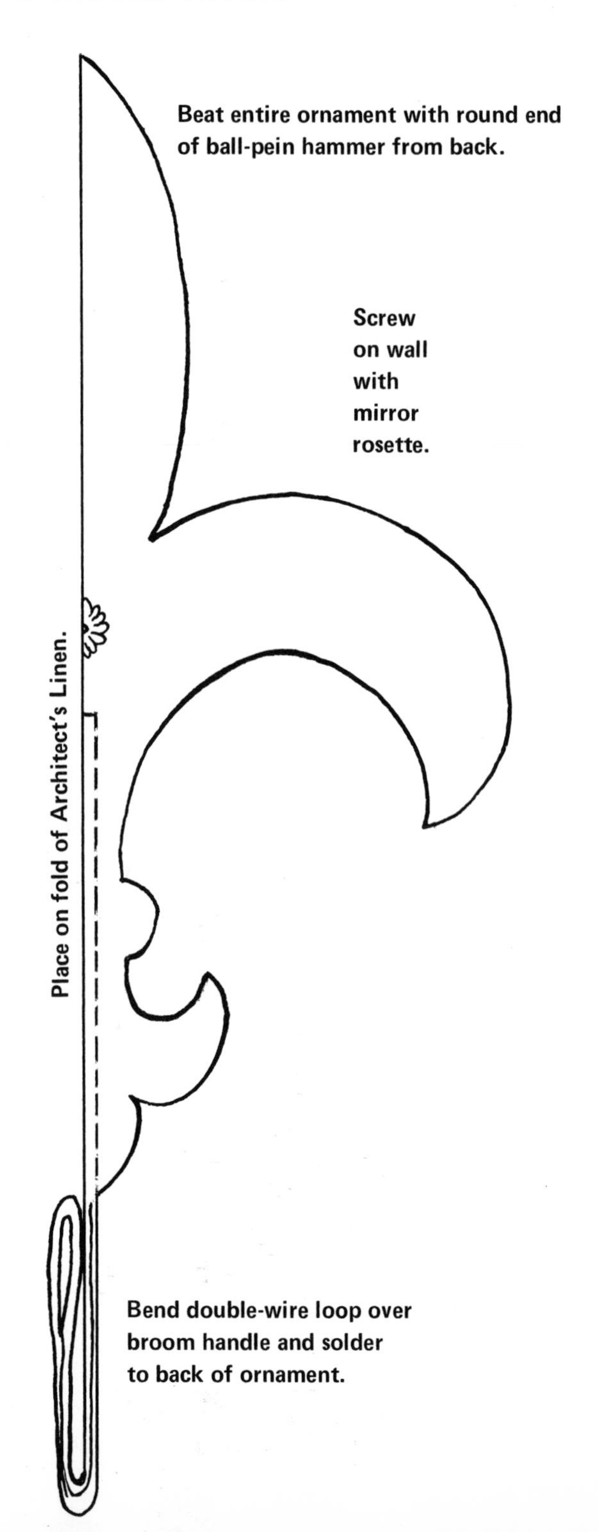

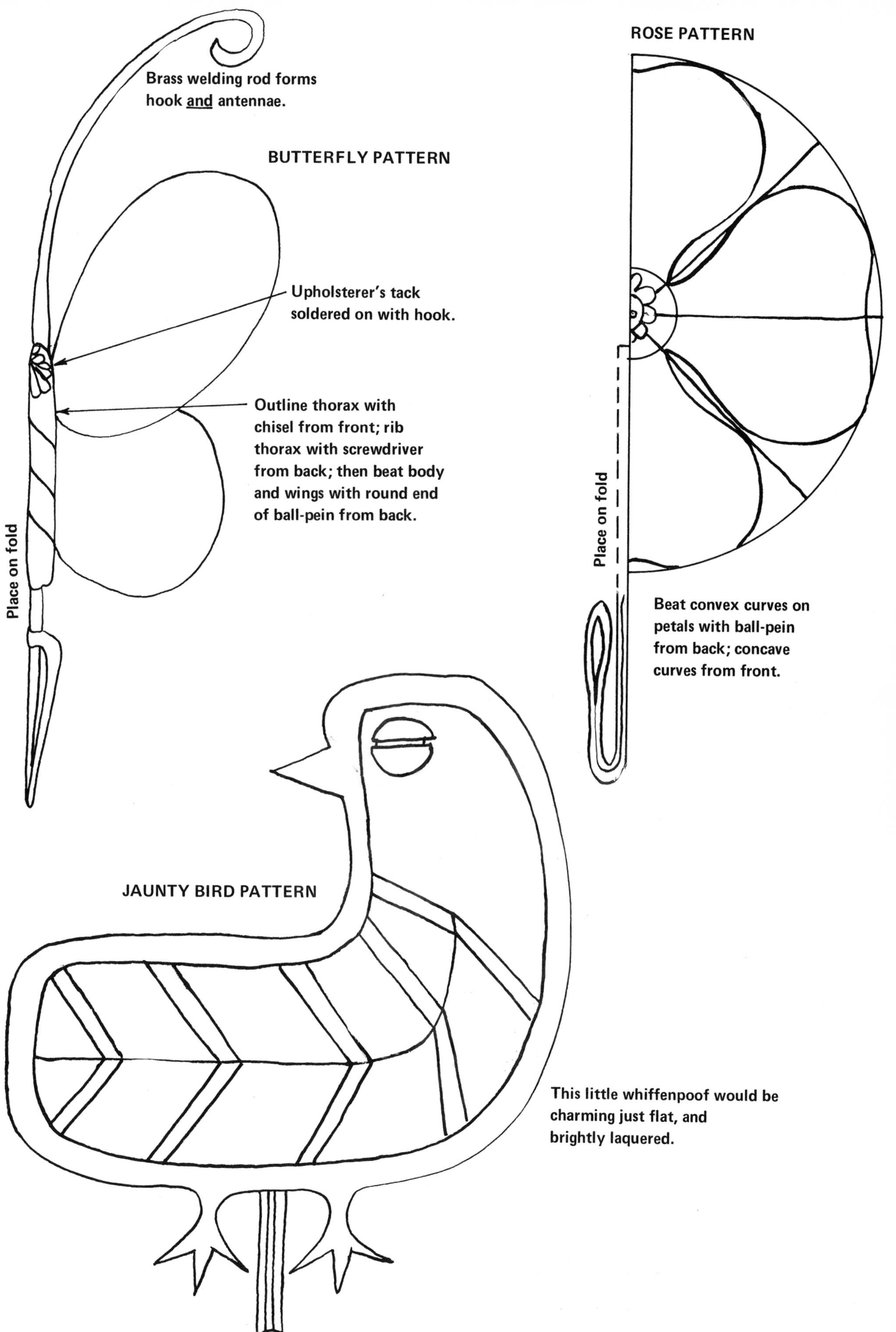
ROSE PATTERN
Brass welding rod forms hook and antennae.
BUTTERFLY PATTERN
Upholsterer's tack soldered on with hook.
Outline thorax with chisel from front; rib thorax with screwdriver from back; then beat body and wings with round end of ball-pein from back.
Place on fold
Place on fold
Beat convex curves on petals with ball-pein from back; concave curves from front.
JAUNTY BIRD PATTERN
This little whiffenpoof would be charming just flat, and brightly laquered.

GREEK BELL KEY RING

Just as pleasingly primitive in aspect as the Greek original, even if not quite so sonorous. It makes a good teething ring too.

TOOLS
Plain-edged snips
Round-nosed pliers
Long-nosed pliers
Awl
Ball-pein hammer and board

MATERIALS
Side of a richly lacquered can
Nu-Gold or florist wire
Architect's Linen, scissors, pencil
Rubber cement
Key ring

Trace the pattern on a fold of Architect's Linen, cut out, and apply to the side of a richly lacquered can with rubber cement. Cut out and hammer flat.

Strike the awl holes for the ring from the top (or gold) side.

Hammer the piece with the round end of the ball-pein hammer on a flat board from the underside.

To make slight sides on your bell, curve the edges with flicks of the long-nosed pliers.

Partially close the bell, grasping it across the middle with the round-nosed pliers. This will ensure a smooth curve instead of a sharp crease.

Cut two wires 1¾" long for the clapper and the loop. With the round-nosed pliers, shape them as shown. Insert the wire loop through the holes in the bell, hook on the clapper, and close the loop.

Squeeze the bell shut so the turned sides partially overlap. Tap the sides with the hammer so they are smooth and flat.

Attach bell to key ring.

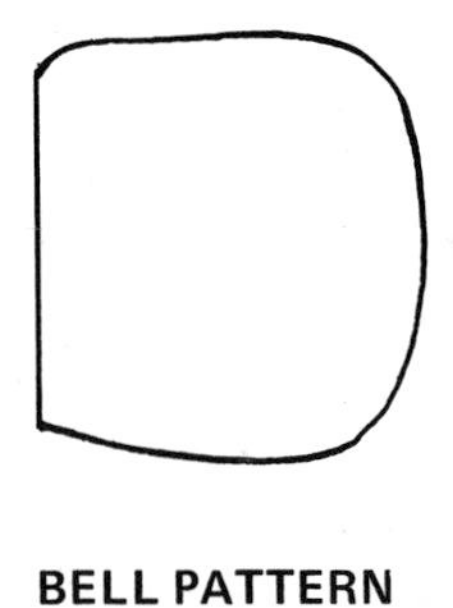

BELL PATTERN

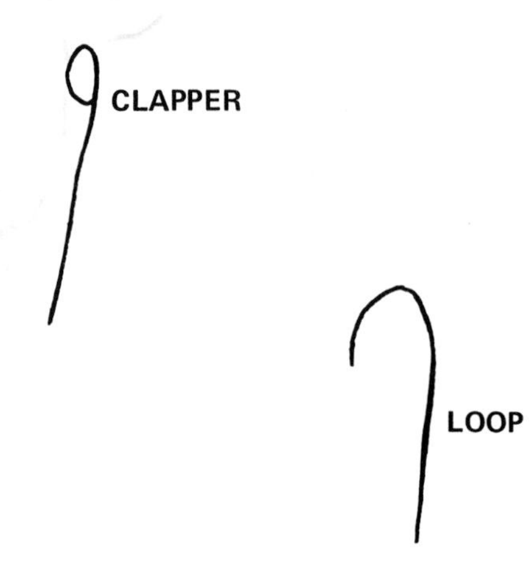

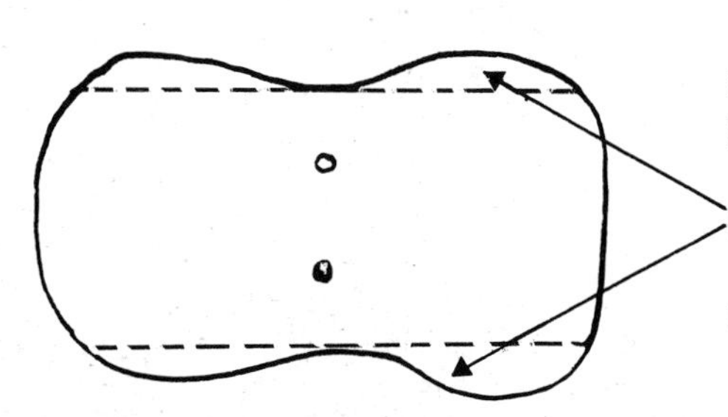

Beat bell from back with round end of ball-pein hammer on flat board.

Curve edges under with long-nosed pliers.

Insert wire loop and clapper before closing bell.

DINING WITH A DIFFERENCE

Unless you'd been indoctrinated by a trip to Mexico, dining with tin utensils might seem decidedly different, although there would be nothing new about it, actually, even here north of the border. At the time our country declared its independence, tin containers were the norm and often the pride of many households, and today, articles hand-fashioned in tin from 1800 to 1850 are valued collector's items.

But after 1850, with the wholesale importation of china, industrial expansion, growing affluence, et al, tinware was largely relegated to the kitchen or camp, and in this century, with the advent of stainless steel and aluminum, it was outmoded almost altogether. Today, however, when our future on this planet appears increasingly precarious, and the need for garnering our resources ever more evident, it only makes sense to think of ingenious ways to reuse containers that come our way. Wherefore, let us dine with a difference—with tin cans.

TIN WAITER An Inspiration

Needing a tray suitable for serving punch on the terrace, Mrs. Alden Vose, with characteristic ingenuity, converted the lid of a large oil drum.

PLACE-CARD HOLDERS

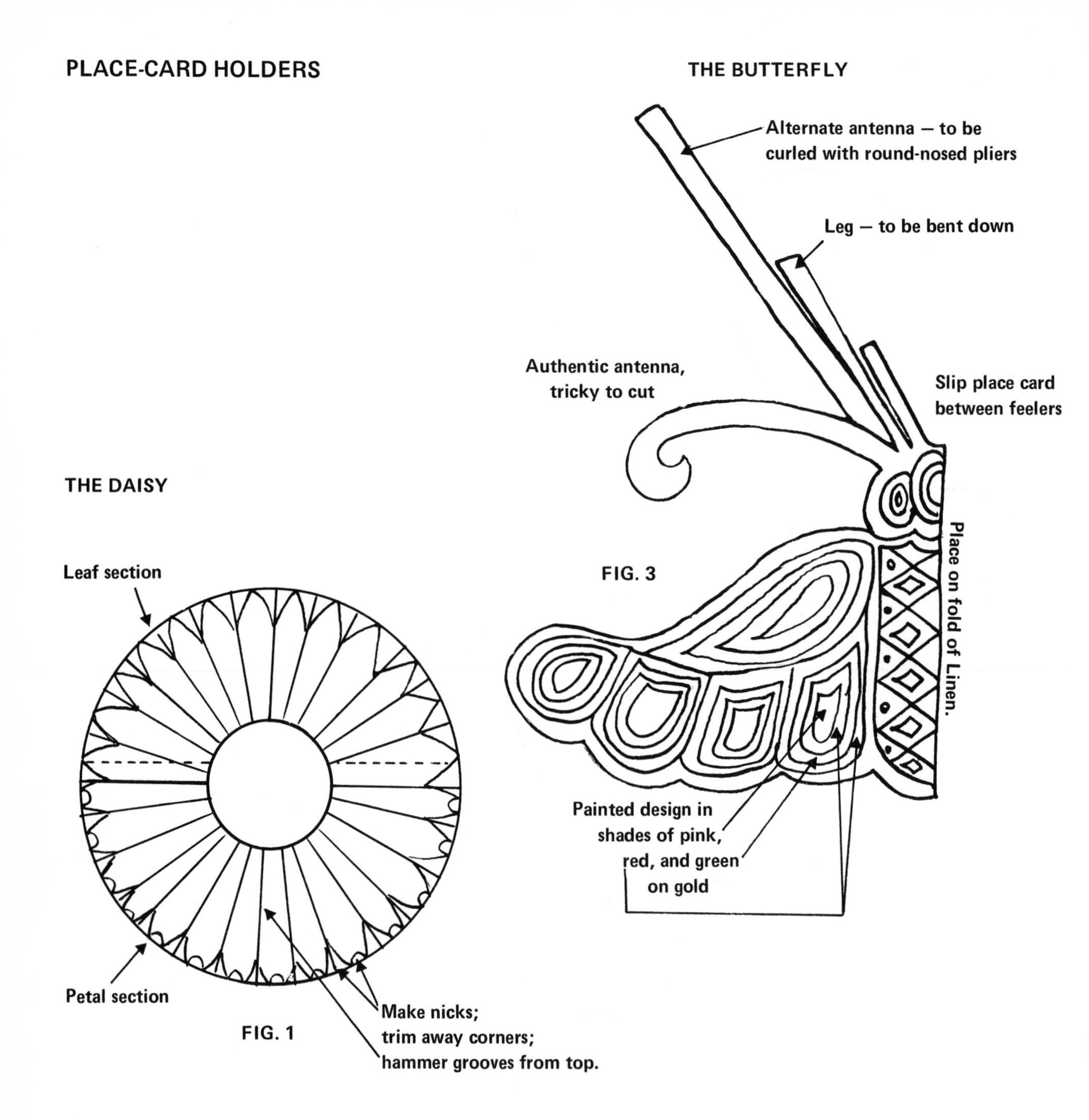

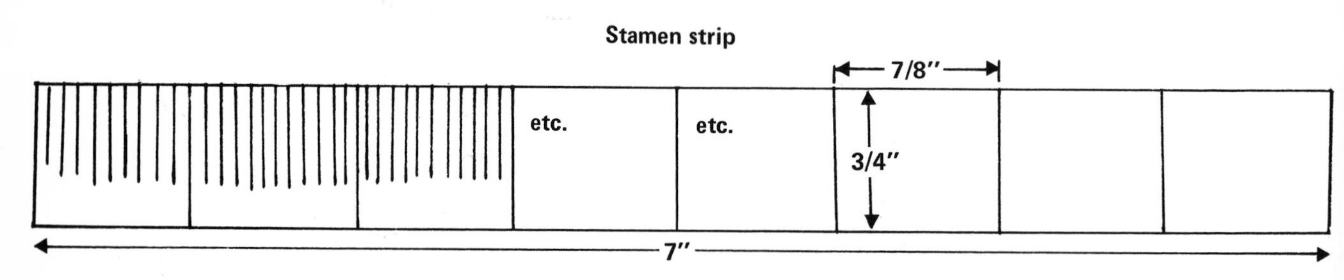

FIG. 2

PLACE-CARD HOLDERS

Of course you can always use a seating chart for your guests and gaily point the way to their places, but it's ever so much more amusing to use place cards, displayed in the petals of a daisy or the antennae of a butterfly. Daisies are so easily cut from the lids of cans that you could make eight, even on the day of the party (if absolutely necessary!). But the butterfly, on the other hand, is a modest work of art which requires a contented sort of puttering with paints and takes planned time. Yours is the choice in meeting the needs of the moment. Perhaps you, too, have daisies in the market all year round, and find their freshness especially appealing.

THE DAISY

TOOLS	MATERIALS
Plain-edged snips or kitchen shears	Lids, 2½" in diameter
	1 unribbed can
Hammer and board	Krylon Flat White Spray Enamel
Chisel	Dri-Mark light green permanent pen
	Tube of yellow acrylic paint
	Brush
	Weldit Cement
	25¢ piece, lead pencil

Select as many lids as you will need and hammer them flat. On the gold side of the lid, place a 25¢ piece in the center and trace around it with a pencil.

Divide the lid into quarters *up to* that center circle (see Dividing Lids Into Equal Sections, page 15). Cut those quarters in half, and cut those halves in thirds (Fig. 1). Hammer flat.

If you have time for an extra touch (and also have good snips, because kitchen shears won't do it), make little nicks in *half* the petal tips and trim off the corners. Hammer flat.

If you don't have the time, shape each petal by trimming away the corners and hammer in the grooves the length of the petals. Hammer these grooves from the top and *do only half the flower*. The other half will be bent down as a base, and be painted green to suggest leaves.

To make the stamens for the daisies (and these will actually hold the place card), cut a strip of tin ¾" x7" from the side of an unribbed can and fringe one side to the depth of ½" (Fig. 2). (This will make eight sets of stamens ⅞" wide, or as wide as the center circle.) While still in one piece, hammer the fringe flat.

With Krylon Flat White Enamel, spray-paint the *silver* side of the daisies and *both* sides of the strip of stamens. It may take two light sprayings to get them snowy white. Be sure they are dry before handling them; they must be pristine!

When perfectly dry, bend the leaf section down at right angles over a counter edge. Then, to form a little stand in that leaf section, bend up the two center leaves (see photograph). With light green Dri-Mark permanent pen, paint the inside of the leaf section, that is, the side that will face you at dinner.

Finish the stamens by cutting the fringed strip into eight ⅞" sections. Separate the fringe, pulling every other strand forward. Glue them onto the back of the leaf sections.

Daub the half-moon center of the daisy with yellow acrylic paint, just as it comes thick from the tube. Tip each stamen with it too. Then, add little green "lashes" around the half-moon.

THE BUTTERFLY

This stylized butterfly was literally lifted from a piece of antique Chinese tea paper; and his broad wings, richly painted, make him a most elegant accessory in a formal table setting. You don't need to paint him, of course; he would be handsome in either beaten gold or silver, but if you do decide to paint him, cue his colors to your china.

TOOLS	MATERIALS
Plain-edged snips	Sides of cans with rich gold linings
Ball-pein hammer and board	Opaque artists' oils
	#0 Fashion Design brush
Round-nosed pliers	Architect's Linen, scissors, pencil
Chisel	Rubber cement

First decide which set of antennae suits your personal wavelength; one, you notice, is harder to cut and requires good Klenk snips.

Trace the pattern onto the dull side of a fold of Architect's Linen (see Tracing Patterns, page 16); cut out; and apply with rubber cement to the flattened side of a can.

With plain-edged snips, negotiate the curves; remove the pattern and the glue; hammer flat.

If you plan to paint him, you may want to draw in the design with pencil first, as a guide.

If you prefer him gold or silver, you will want to delineate the body and wing sections with a chisel, and beat both the wings and the body with the round end of the ball-pein hammer on a flat board.

Curl the alternate antennae with the round-nosed pliers. Bend down the legs. The place card will slip between the stubby feelers and the wavy antennae.

COCKTAIL PICKS

Charming cocktail picks are almost impossible to find, and anyway, it's more fun to make your own. Leaves are the easiest, and then the little tulips. Enameled with Talens Glass Paints, they will look to your guests like French imports. How about selling them at the fair?

TOOLS

Plain-edged snips
Ball-pein hammer and board
Metal hammering block
Screwdriver
Propane torch
Asbestos mat

MATERIALS

Sides of cans
Nu-Gold or 20-24 gauge florist wire
Solder, Dunton's Tinner's Fluid
Talens Transparent Glass Paints
Steel wool

From the side of a can, cut a collection of leaves or blossoms, whatever amuses you and is simple in outline. Hammer them smooth. Trim any unevenness and hammer again.

Cut 3″ lengths of Nu-Gold or florist wire (or silver wire; see Sources of Supply, page 195) and hammer both ends flat on a metal hammering block, or any smooth hard surface.

Solder the wires onto the flowers (see Soldering, page 20). Rinse them in water and burnish with steel wool.

Enamel them richly with Talens Transparent Glass Paints.

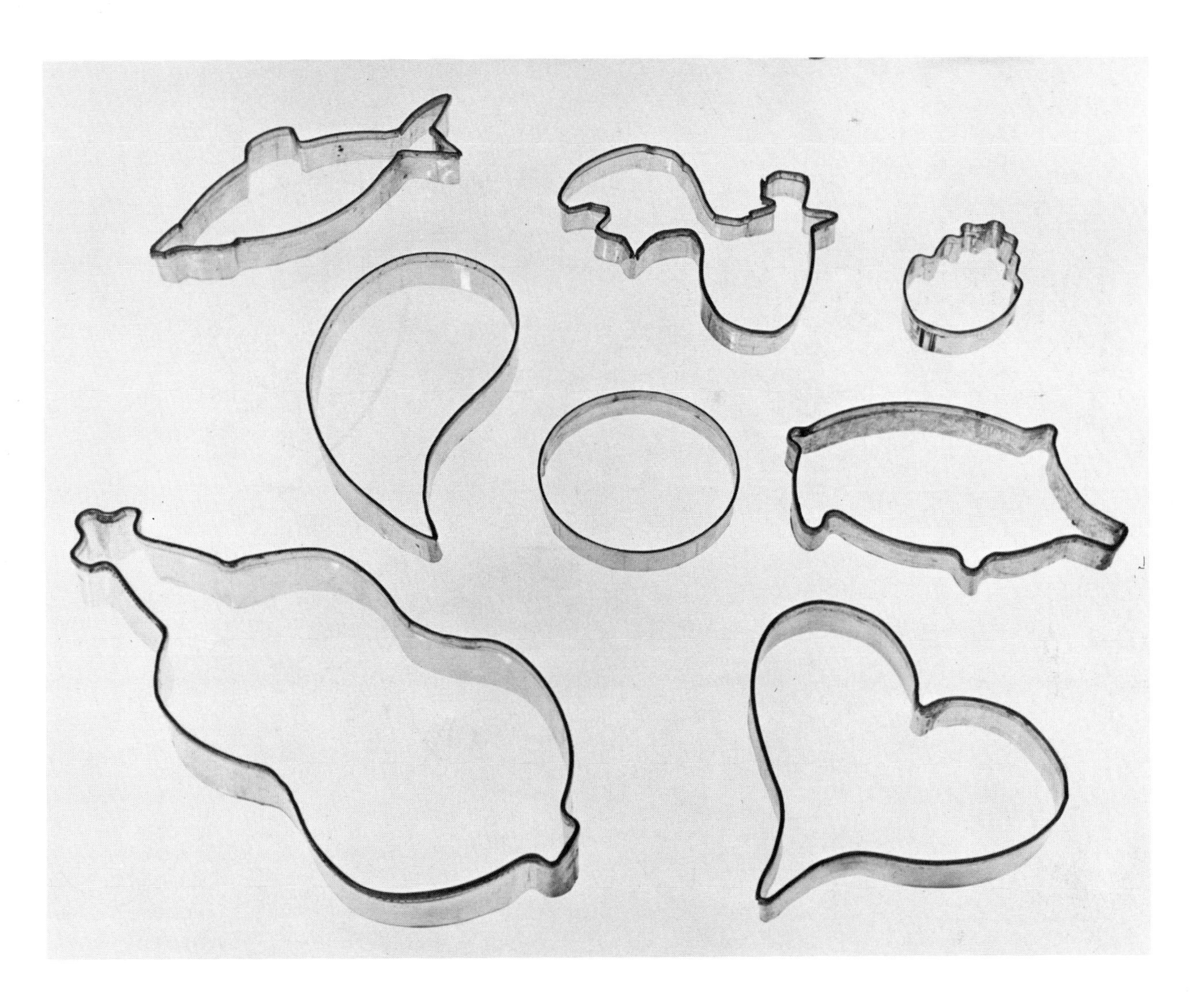

KOOKY CUTTERS

Somehow it's twice as much fun to eat a Gingerbread Man as it is a gingerbread cookie. And the same with a Parmesan Pig and a Cardamom Cock.

All you need for an instant cookie cutter is a cut-down can large enough to go around all the humps and bumps, plus a pair of long-nosed pliers. The shapes should be expressive but simple. You could draw a pattern and use it as a guide, but it's largely a matter of judgment that makes it all come out even in the end.

After carefully cutting down the can to ½″ or so from the bottom rim (see Cutting Down Cans, page 14), start shaping the cutter at the point in the design that interests you most, like the nose of the pig, or the tip of the leaf, or the peg box of the guitar. Grasp the rim along the seam with the pliers, and together with your helping hand, twist the pliers and bend the strip in the direction it should go. Move on to the next point of interest and make the necessary bends there, etcetera.

After bending the rim into shape, true up the *cutting* edge to make sure that it *corresponds to the shape of the rim* above it, and is perfectly *perpendicular to it;* otherwise it won't cut through the dough evenly. (To keep it true, you could solder a flat piece of tin onto the rim; perhaps even Epoxy 220 would hold it.)

These cookie cutters would never suit a perfectionist; they are just for fun. But you'll find that your original shapes will entertain your guests as much as they did you in the making.

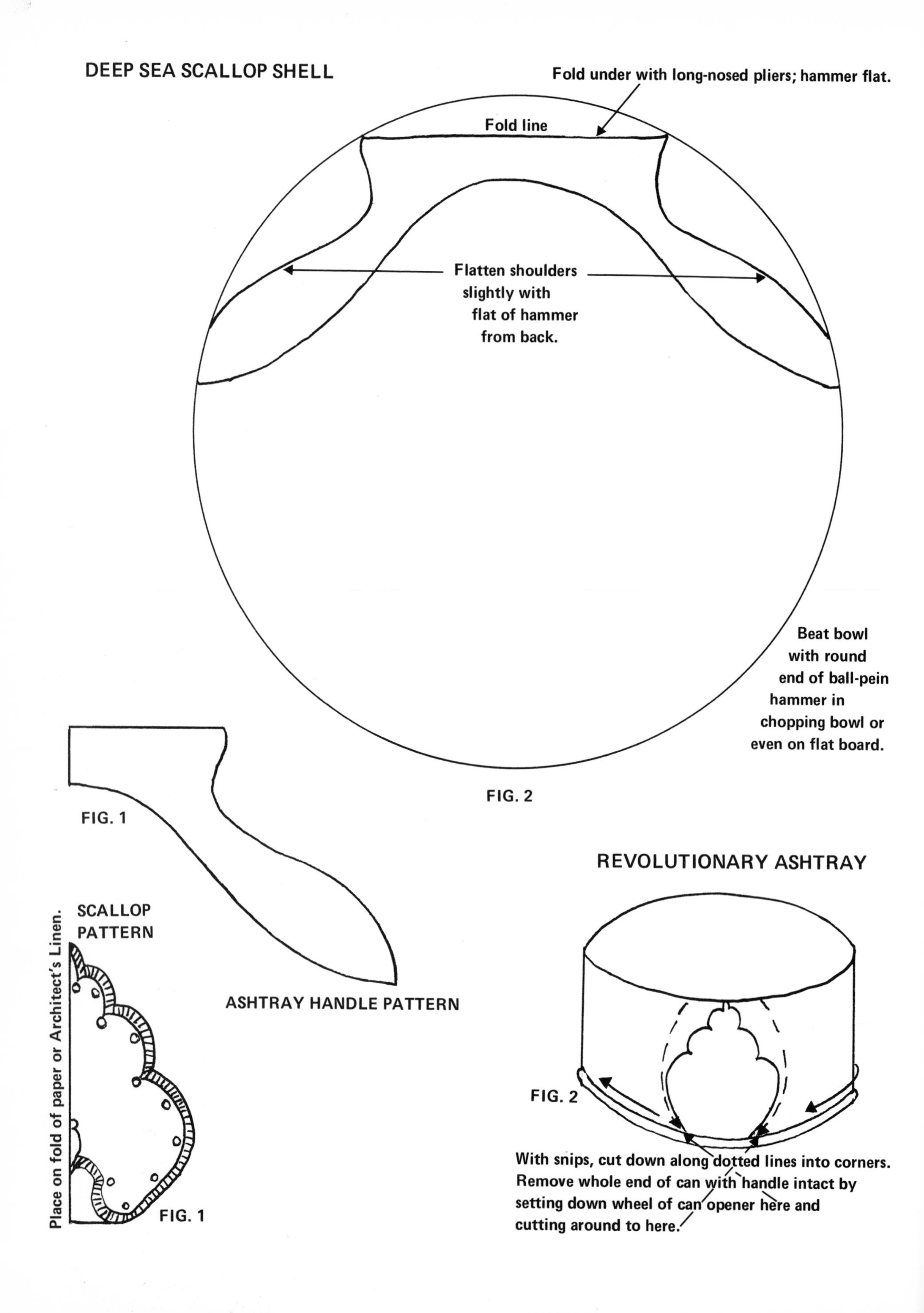
DEEP SEA SCALLOP SHELL
Fold under with long-nosed pliers; hammer flat.
Fold line
Flatten shoulders slightly with flat of hammer from back.
Beat bowl with round end of ball-pein hammer in chopping bowl or even on flat board.
FIG. 2
FIG. 1
ASHTRAY HANDLE PATTERN
SCALLOP PATTERN
Place on fold of paper or Architect's Linen.
FIG. 1
REVOLUTIONARY ASHTRAY
FIG. 2
With snips, cut down along dotted lines into corners. Remove whole end of can with handle intact by setting down wheel of can opener here and cutting around to here.

DEEP SEA SCALLOP SHELL

There's something about a scallop shell, shallow and symmetrical, lying there open and ready for use. What is that line of Raleigh's? "Give me my scallop shell of quiet . . ."? But aside from poetic associations, it's amusing to me, deep in the Maine woods in summer, to prepare a Mountain Pond version of Coquilles St. Jacques made from tuna fish, served in scallop shells fashioned from the lids of the tuna cans.

TOOLS

Plain-edged snips or kitchen shears
Ball-pein hammer and board
Long-nosed pliers
Chopping bowl

MATERIALS

Lids from family-pack tuna cans or 2-pound coffee cans
Any paper for pattern, pencil
Rubber cement

Any lid 4¾″ in diameter will make a good, average-size scallop for an individual cocktail serving of creamed, crumbed fish. By enlarging the pattern, you could even make a huge scallop shell from the lid of an enormous shore-dinner can. All it would require is a little persistence; do it while you sun on the dock and watch the children fishing for sunnies.

Hammer the lids flat. Trace and cut out pattern (Fig. 1); open pattern and apply with rubber cement to the lid. Use first shell as pattern for others.

Bend the handle under as indicated, and beat the bowl of the scallop with the round end of the ball-pein hammer in your kitchen chopping bowl, until it is nicely rounded.

REVOLUTIONARY ASHTRAY

No doubt the scallop shell would be used as an ashtray if you left it around, but here is another design, reminiscent of Paul Revere, that is *really* meant to be an ashtray.

TOOLS

Plain-edged snips
Ball-pein hammer and board
Awl
Filed-down nail or tiny screwdriver
Oval die, or sawed-off screwdriver

MATERIALS

Family-pack tuna can (or something similar)
Architect's Linen, scissors, pencil
Rubber cement

Remove the top rim of the can.

Apply the handle pattern with rubber cement to the side of the can, so that the base of the pattern is flush with the bottom rim (Fig. 1).

With plain-edged snips cut down from the top of the can into the corners where the pattern meets the rim.

Placing the can under the wheel of the can opener sidewise (see Removing Rims, page 14), cut off the sides of the can, leaving the handle intact.

Gently bend the handle out. Carefully cut around the pattern (see Cutting Out Patterned Designs, page 16). Remove the pattern, and refine the curves. (Unless you know how to solder, treat the handle with respect lest it be tempted to part company with the body.)

Beat the bowl of the ashtray with the *flat* end of the ball-pein on a flat board to remove the circular ridges in the end of the can. Then, tilting the bowl at a slight angle, beat around the inside of the rim with the round end of the ball-pein hammer. You will not need to beat the very center of the bowl at all, since you want the ashtray to rest level on the table; it's just that inch around the edge that you need to make concave.

Ornament the handle and the bowl any way you wish, possibly using some of the tools listed.

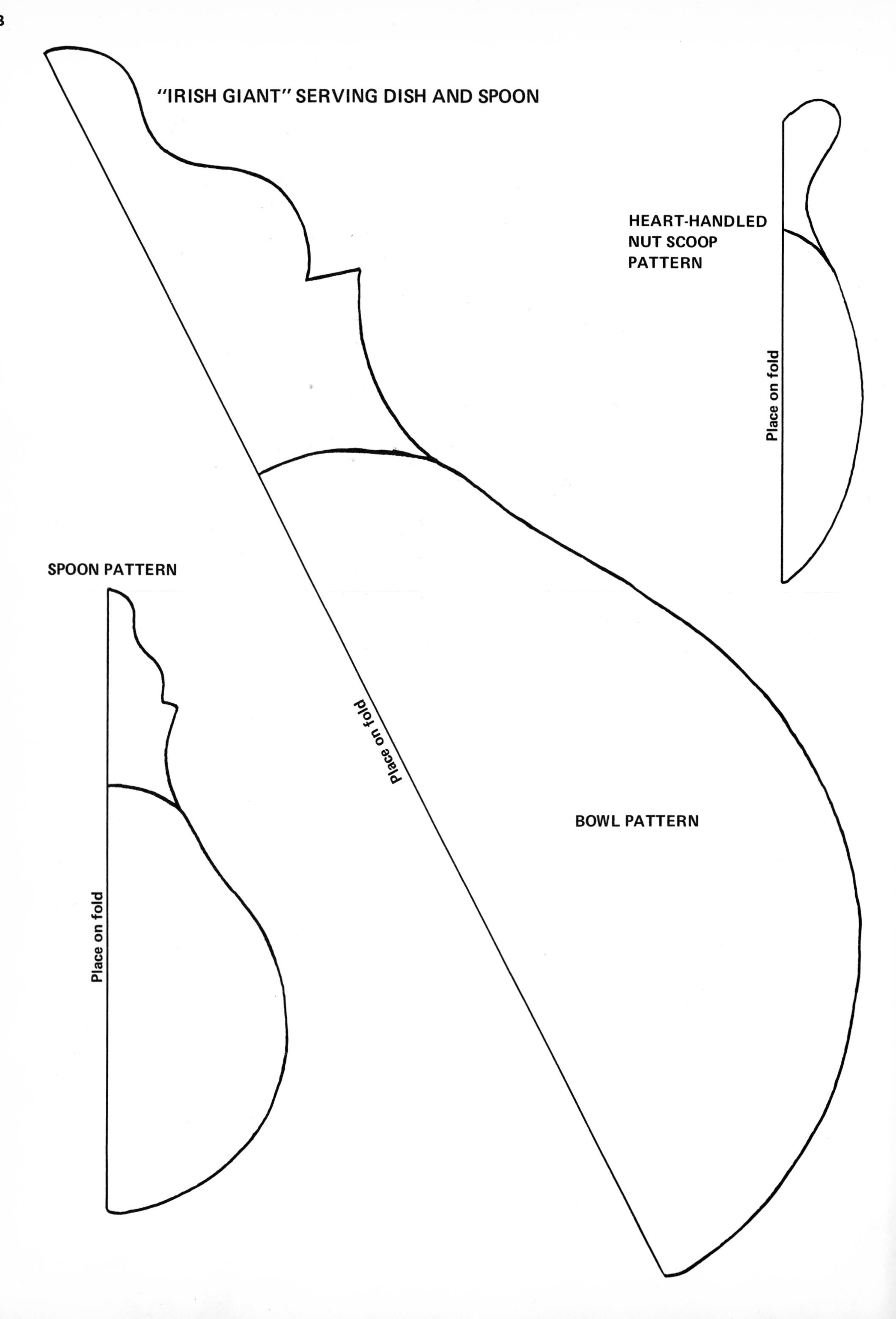
"IRISH GIANT" SERVING DISH AND SPOON
HEART-HANDLED
NUT SCOOP
PATTERN
Place on fold
SPOON PATTERN
Place on fold
Place on fold
BOWL PATTERN

"IRISH GIANT" SERVING DISH AND SPOON

This tea caddy spoon, designed in 1815 by a Dublin silversmith, was called the "Irish Giant" because of its exceptional size, measuring five inches by three inches. It would be nice as a scoop for nuts, crystallized lilac blossoms, whipped cream, or whatever. Have some fun and design your own escutcheon for the handle, or even leave it plain, since the shape has considerable distinction.

TOOLS
Plain-edged snips
Ball-pein hammer and board
Assorted nails for ornamenting
Chopping bowl

MATERIALS
Side of an unribbed can, preferably imported because of lighter weight
1 soup can rim
Epoxy 220

Making a bowl from a flat piece of metal is chiefly an exercise in curiosity and patience, in the end rewarded with a quiet sense of satisfaction. It also requires what one author described as a "profound intimacy between the craftsman and his material." But don't let this preamble put you off. Start with the spoon, working in the bottom of the chopping bowl with the round end of the ball-pein hammer, making many little hammer strokes back and forth around the edge.

Very gradually work toward the center. Rest your arm every minute or so. Go away, if necessary, and come back; but whatever you do, don't hurry it or you'll tear the metal. Very patiently, lots of little hammer strokes. That's all.

Glue a soup can rim to the bottom of the serving dish so that it will sit properly on the table.

My hand-turned rim is a little crude; see if you can't do it better. I used the long-nosed pliers, flicking them around the edge to flare it; then I hammered the lip with the flat of the hammer on the very corner of the board so as not to put bends in the can circle.

The saucer for the ramekin is the ornamented lid of a 2-pound coffee can.

Speaking of ornamentation, do as much or as little as you like, drawing the design on with felt-tip pen, if that makes it easier. The serrated snips add a decorative effect, but make the edge a little rough to the touch. Whatever you do, be sure to hammer the edge well.

GADROONED RAMEKIN

You could serve anything from shrimp cocktail to caramel custard in this ramekin.

TOOLS
Plain-edged snips
Long-nosed pliers
Ball-pein hammer and board

MATERIALS
Unlacquered can about 3¼″ wide
Lid from 2-pound coffee can
Shelf paper or Architect's Linen, scissors, pencil
Rubber cement
Ruler and felt-tip pen

You can see that the ramekin is a cut-down can with the handles and bowl all one piece. The easy way to make sure the handles are opposite each other is to cut shelf paper (or Architect's Linen, if making several) to fit around the can so the ends butt. Fold the paper in half and crease it well. Open it up and fold both ends to meet at the crease.

Rule off the rim-line 1″ up from the bottom, and trace the handle pattern onto both folds (Fig. 1). Cut out your choice of handle patterns (Fig. 2 or 3) and apply it to the side of the can with rubber cement so that the ends meet at the seam.

In cutting out the leaf handle, bypass the serrations and make a smooth curve down to the corner where the handle joins the rim-line, first on the right side, then on the left. When you reach the rim-line on the left, about-face your snips, right there in the corner, and, proceeding in a westerly direction, cut along the rim-line to the other leaf. Do the same with the second leaf.

RAMEKIN PATTERN, FOR PROPER PLACEMENT OF HANDLE

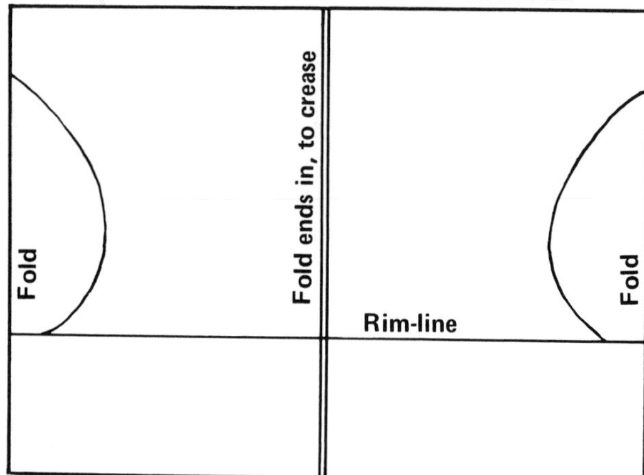

FIG. 1 Cut piece of shelf paper (or Architect's Linen) to butt around can.
Fold in half and crease.
Open up and fold ends in, to center.
Rule rim-line 1″ up from bottom.
Trace handle pattern on both folds.
Cut out and apply to can with rubber cement.

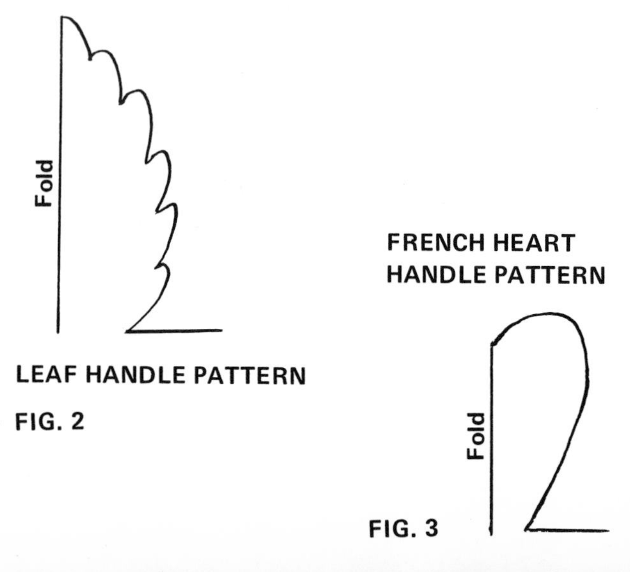

LEAF HANDLE PATTERN

FIG. 2

FRENCH HEART HANDLE PATTERN

FIG. 3

PART FIVE
confections

FOR THOSE SPECIAL OCCASIONS

For me, all occasions are special, if not by nature, then by invention. For what better way is there to celebrate the gift of life and the joys of friendship than conjuring up a little confection to heighten one's pleasure in the moment?

SENTIMENTALIA A Collage

Designed to bridge the gap between home and prep school, this is a collection of gear dear to the heart of our "new boy." Here is "home" in a fresh context, to keep him company in the cold corridors of academia.

VALENTINES

Should you need extra incentive, Valentine's Day provides the perfect opportunity to send greetings to special friends, and half the fun is in devising some whimsical way of expressing your affection . . . something slight but surprising . . . just to let them know you are thinking of them.

TWO KINDS OF LOVE

Of course I wanted to send love to my two youngest, who were off at school, and that meant that whatever slight thing I thought of should be able to slip into an envelope for mailing and hang on the wall once there. Wherefore, the two tin "LOVE" signs in the color photograph on page 68. My daughter's is made of one piece of tin, beaten with the ball-pein hammer, painted with Dri-Mark pens, and mounted with butterflies on a yellow plastic cloud nine.

My son's consisted of similar, but separate, letters, glued together and surrounded by abstract curves and spirally wires. In the O of "LOVE" was an eye, as if to say, "Here's looking at you!" And after the E was a question mark made with a cup hook, as if to say, "Guess who?" Just one of a zillion ways to say "love."

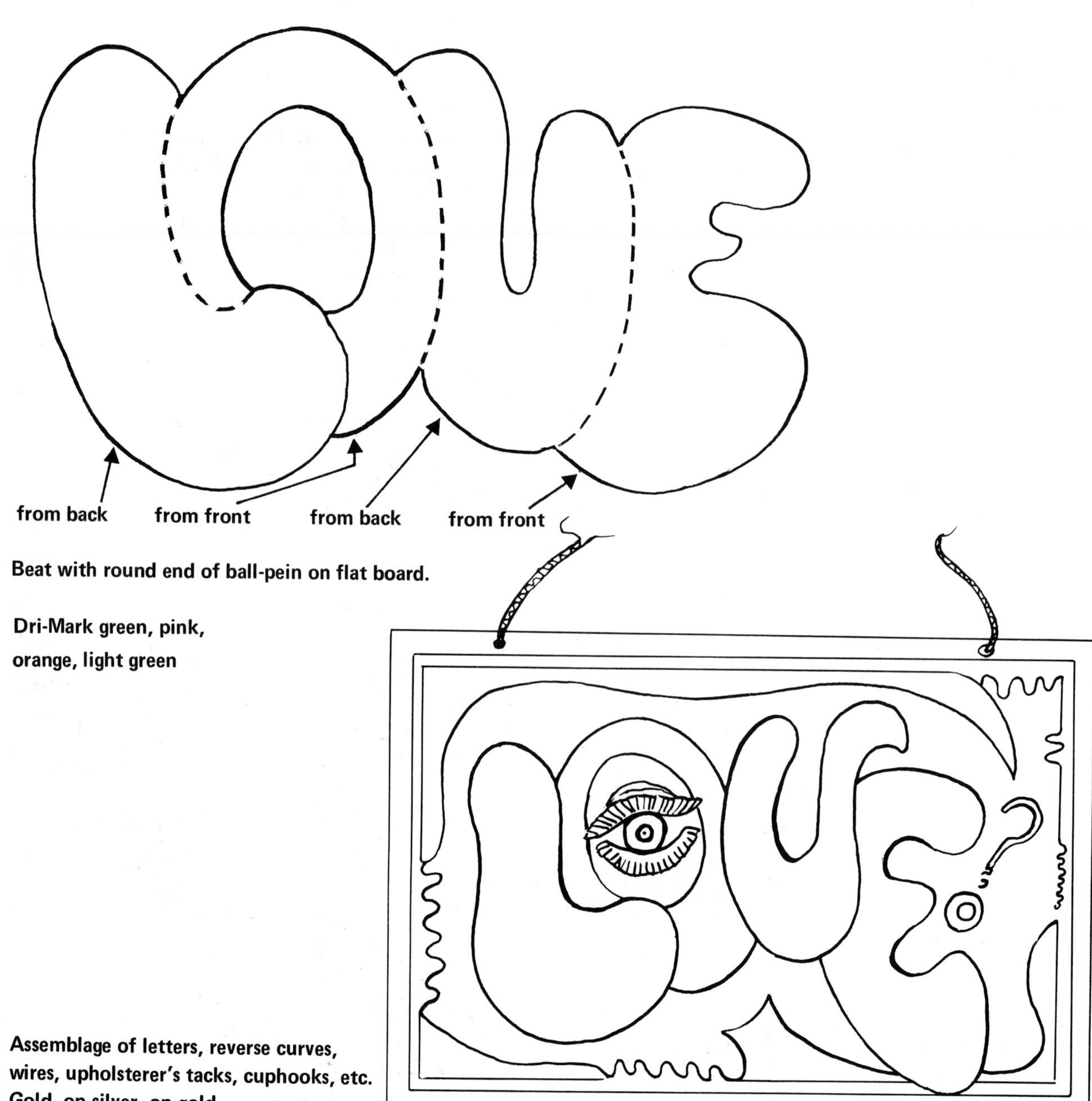

Beat with round end of ball-pein on flat board.

Dri-Mark green, pink, orange, light green

Assemblage of letters, reverse curves, wires, upholsterer's tacks, cuphooks, etc. Gold, on silver, on gold.

LITTLE PINK PIG

If you detect a Picasso-esque air about the little pink pig in the color photo on page 68, it is because my friend Arthur Train, who drew it, had just finished writing a book about Picasso. The piggy was sketched in pink, sky-blue, and leaf-green pastels, within a ruby-rimmed heart that was inscribed, "Why be a . . . when you can be my Valentine?"—a delightful bit of nonsense which I have enjoyed so much I decided to share it with you. The pig makes a cunning Christmas tree ornament also, painted white with sprigs of holly.

TOOLS

Plain-edged snips
Hammer and board

MATERIALS

Side of an unribbed can
Tracing paper or Architect's Linen, scissors, pencil
Rubber cement
Carbon paper *(optional)*
Krylon Hot Pink Enamel *(optional)*
Artists' oils, #0 Fashion Design brush, varnish *(optional)*
Dri-Mark felt-tip pens *(optional)*

If you are going to make several piglets, use Architect's Linen for the pattern, tracing both the outline and as much of the flower design as you need.

Cut out the pattern and apply it with rubber cement to the tin. With plain-edged snips, cut out the pig, remove the pattern, and hammer him flat.

Unless you plan to use Dri-Mark pens on the plain tin instead, spray-paint the pig on both sides with Krylon Hot Pink Enamel.

Slip a piece of carbon paper between the pattern and the pig, and trace on the design with a ballpoint pen.

Paint the flowers with artists' oils diluted substantially with varnish.

Cut his corkscrew tail.

Color key: Spray-paint piggy shocking pink
Flowers: Orange with white dots
Leaves: Light green with blue veins.
Eye: Royal blue outline and pupil; Green "eyeshadow"; White iris.
Ears: Blue outline

LITTLE PINK PIG PATTERN

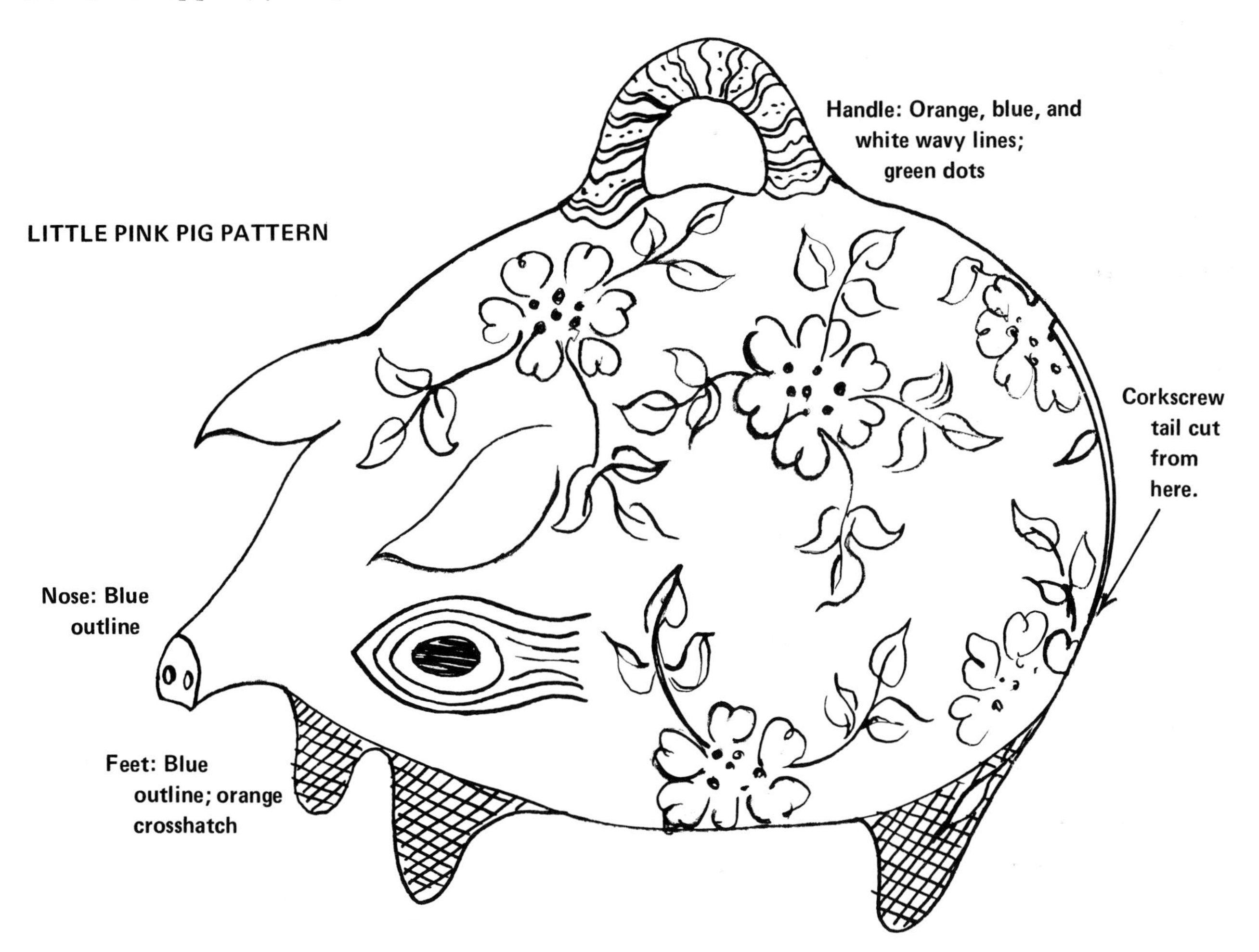

HEART KEEPSAKE

Apparently during the latter half of the fifteenth century there was a vogue for heart-shaped illuminated manuscripts. The one which inspired this keepsake (pictured in the color photo on page 68) was unique in that it is the only known surviving manuscript in the shape of one heart when closed, and two, joined hearts when open. The original was made in 1470 for a naughty French cleric named Jean de Montchenu, and contains lilting love songs of the period, but I can think of a dozen loving purposes to put this twentieth-century version to, can't you? It could be a graduation gift for your daughter's classmates' autographs. Or an engagement present for a bride-to-be in which she could list all her presents, or register the guests at her wedding reception. It could be a valentine for your true love with your own poems carefully penned. Or a baby book. Or a photograph album for a special occasion. Or what you will.

TOOLS

Snips
Ball-pein hammer and board
Awl
Round-nosed pliers

MATERIALS

Approx. ¼ yard velvet or brocade
Cardboard, plain or corrugated
Shelf paper, scissors (or Exacto knife), pencil
Rubber cement
Gift-wrap paper or other endpaper
Parchment or rice paper for pages
Embroidery thread
Metallic thread, Weldit Cement, toothpick
Piece of unribbed tin approx. 10″ long
and 6″ wide for hinge and clasp
Nu-Gold or florist wire

THE BOOK

Trace the heart pattern on a fold of shelf paper (Fig.1). Cut out, open up, and apply to cardboard with a few dots of rubber cement, just enough to hold it in place while you trace around the double heart.

Cut out with scissors or Exacto knife.

If you are going to have only a *few* pages in the book, *score* the cardboard along the fold line with awl or Exacto knife.

From the same pattern, cut pages for the book out of sheets of parchment (or whatever) and sew them into the fold line with embroidery thread. (You could also use the basting stitch on your sewing machine, if you prefer.)

However, if you are going to have quite a number of pages in the book, with photographs on each, which would tend to make the volume thick, then you will have to *cut* along the fold line on the cardboard and treat the whole thing, pages and covers, as you would a proper book. Study a commercially bound book to see how groups of pages are sewn together and then joined to a cloth backing. Notice how the cloth binding on the covers is glued to the back of the stitched pages, etcetera. It's not difficult (I bound a book in tooled leather when a camper of thirteen), but it takes time and cannot be done on the spur of the moment, which is the way most of us do things these days.

THE BINDING

Assuming that you intend to make a small and precious volume, and have sewn your pages into the backing, proceed to bind them.

Iron the material you plan to use (if it needs it), and lay it flat on a table.

Coat the cardboard double heart with rubber cement and carefully position it on the material. Press down. Allowing ¼″ margin all the way around, cut out the velvet double heart.

Apply rubber cement to the margin *along the straight edges* of the heart, fold them over onto the cardboard, and press.

Work around the curves, noticing where the material bunches and notches need to be cut. Glue and press as you go. Thank heaven, rubber cement comes off most anything, even velvet, if given a moment to dry.

Cut out endpapers from appropriate gift-wrap paper or whatever, using the original heart pattern. Cut in two along the fold line. Try the pieces on for size. If the material you have used for the binding has no nap, and therefore does not increase the overall size of the book, you may want to pare down the outline of the endpapers a bit. But, if you have used velvet, the original heart pattern should just cover the glue-job nicely and leave 1⁄16″ margin. Glue the endpapers in place with rubber cement.

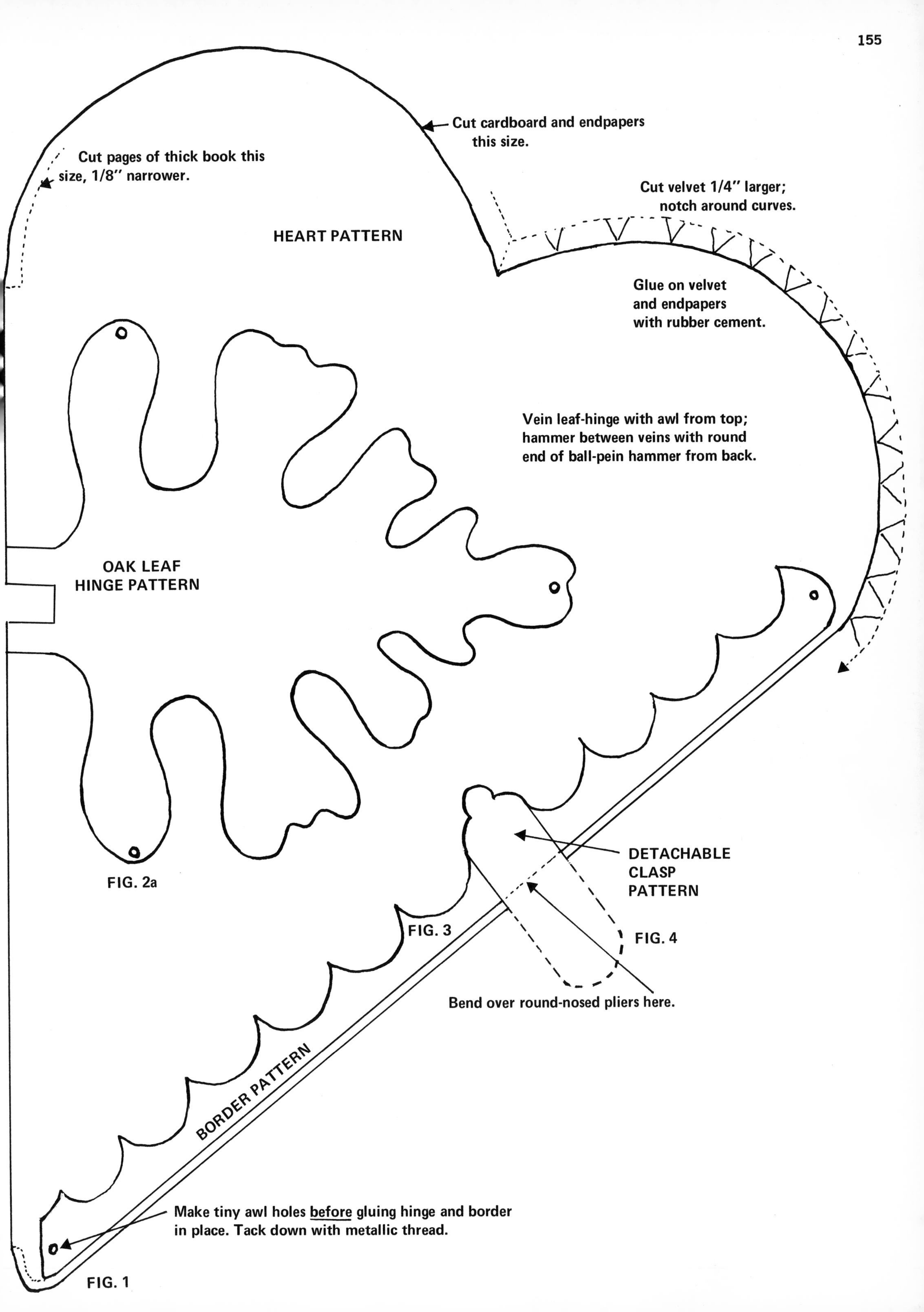
Cut cardboard and endpapers this size.
Cut pages of thick book this size, 1/8" narrower.
HEART PATTERN
Cut velvet 1/4" larger; notch around curves.
Glue on velvet and endpapers with rubber cement.
Vein leaf-hinge with awl from top; hammer between veins with round end of ball-pein hammer from back.
OAK LEAF HINGE PATTERN
FIG. 2a
DETACHABLE CLASP PATTERN
FIG. 3
FIG. 4
Bend over round-nosed pliers here.
BORDER PATTERN
Make tiny awl holes before gluing hinge and border in place. Tack down with metallic thread.
FIG. 1

THE LEAF-HINGE

I always save curious reverse curves that result from cutting out a design, or any pieces that don't fit into a design as I had expected them to, because I know that sometime they will be just right somewhere. So it was with this leaf-hinge and border. The leaf was designed for the fanlight over my front door, but seemed too much there, and the border was meant to encircle a potpourri jar, but was too short. Both were too decorative to be thrown out, so here they are looking quite intentional, don't you think? If you like this leaf design, apply the pattern with rubber cement to the flattened piece of unribbed, polished tin. Cut out; remove the pattern, and hammer the leaf flat. Perfect the outline and hammer flat again. Cut the hinge in two along the fold line.

Score in the veins with the awl on the top side of the leaves (see Veining, page 16). Turn the leaves over and beat in between the veins with the round end of the ball-pein hammer on a flat board.

Make the tiny awl holes indicated in Fig. 2A. Cut the working part of the hinge (Fig. 2B). This is a very rudimentary tongue and groove design, as you can see. Form the little cylinders, through which the wire will pass, just as straight and true as you can with your round-nosed pliers. You may have to squeeze them shut a little tighter with the long-nosed pliers.

Make the pin of Nu-Gold wire, if you have any. Otherwise, your heaviest florist wire will do. Pass it through and curl the tips with the round-nosed pliers as snugly as possible.

Having assembled the hinge, decide where you want to place it along the spine of the book. Mark the place in such a way that it won't show when the hinge is glued on. Apply rubber cement to the inside of the hinge and carefully press it in position. Use your iron to weight the hinge until dry.

When dry, tack the hinge down with double knots in metallic thread through the awl holes. Apply Weldit Cement to the knots with a toothpick.

THE BORDER AND CLASP

Cut the border (Fig. 3) from a flat strip of tin. Tracing around a dime will help you get the scallop even. Strike the tiny awl holes at each end and hammer the strip from both sides until it lies perfectly flat.

Apply rubber cement to the back of the strip and press it in place. Let dry; then tack with metallic thread.

Because the book doesn't quite want to close, fashion a simple clasp (Fig. 4) from a scrap of tin and bend it over the round-nosed pliers until it fits the book properly. (If its being loose bothers you, attach it to the hinge with a length of fine silver chain salvaged from an old necklace.)

Speaking of chains, did you know that for ages books were so valuable they were chained to the reading desks? Now those same books are so valuable that they are kept under glass. Who knows, your little keepsake may wind up under glass, too, one day.

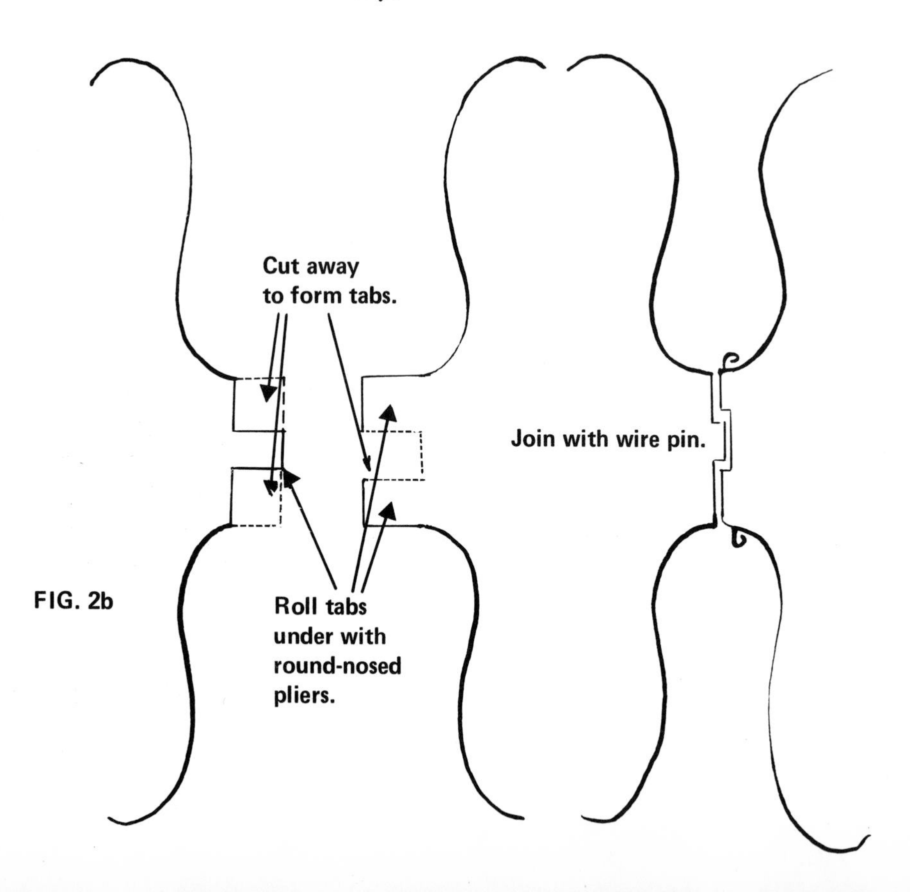

FIG. 2b

BOOKMARKS

Adaptations of Empress Josephine's elaborate fan-sticks, these bookmarks are tucked in here as a gift suggestion, a little plus to add to a package. Mounted rosettes would be pretty. Butterflies too?

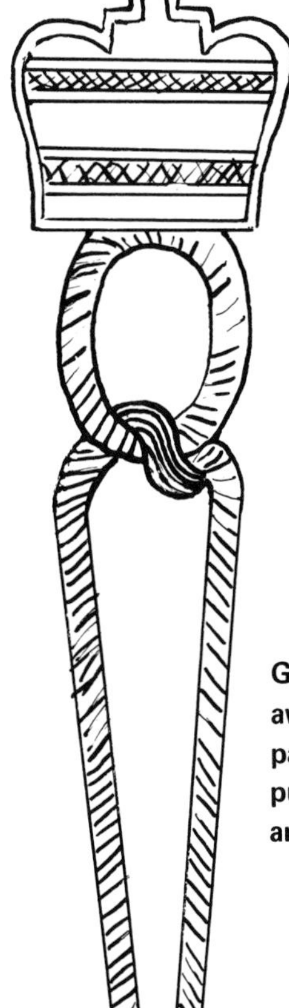

Gold ground, with awl hatching, painted in royal purple, grass green, and sky blue.

TOOLS
Plain-edged snips
Awl
Ruler

MATERIALS
Sides of unribbed, gold-lined cans
Talens Transparent Glass Paints, or Dri-Mark pens
Tracing paper, scissors, pencil
Rubber cement
Krylon Crystal Clear Acrylic Spray *(optional)*

Trace off the design, apply with rubber cement to a choice piece of gold-lacquered tin, and cut out. Hammer flat.

With awl and ruler, mark in the crosshatch. Paint to please yourself. The Talens Glass Paints are as rich in color as real enameling, and when baked in a slow oven, are virtually indestructible. But the Dri-Mark pens (permanent), while less rich, are very gay in color and easier to apply. Let dry over-night and spray lightly with Krylon Crystal Clear Acrylic for added protection.

Gold ground with shocking-pink star, sky-blue flowers, leaf-green border.

Use Talens Glass Paints, baked in slow oven; or Dri-Mark oil-based pens, sprayed with clear acrylic.

THE DRESS

This dress is an example of the lengths to which some parents will go to convince their children they aren't square! However, I suggest you settle for an apron and cap. That would be very funny and much easier.

For a dress you really need a dress form, built to your specifications. And if you want an open cage like this, you need an electric drill to make the holes in the rims for the links, so that it will keep its shape and stand out stiffy-starchy even when you're not in it. If you have ever studied Paco Rabanne's creations, you have probably noticed that he links the costume together through holes punched north, south, east, and west in each individual piece. You can do this with hammer and awl in the lids or sides of cans, but you need an electric drill to penetrate rims.

Even though this dress needs only to be hooked together in the back *from the waist up*—so that it *is* possible to sit down in it and even drive a car—still, it is not what one would call comfortable except when standing. An apron would eliminate the sitting problem altogether, and, of course, could be worn over any dress with a normal waistline, so that you wouldn't have to make a special shift. Furthermore, there's the time factor. You could make an amusing fringed apron and cap in a morning, if pressed, whereas a dress will take you more nearly a week. Not that time should matter on these very special occasions!

THE HAT A Creation

What would *you* do with an old window screen?

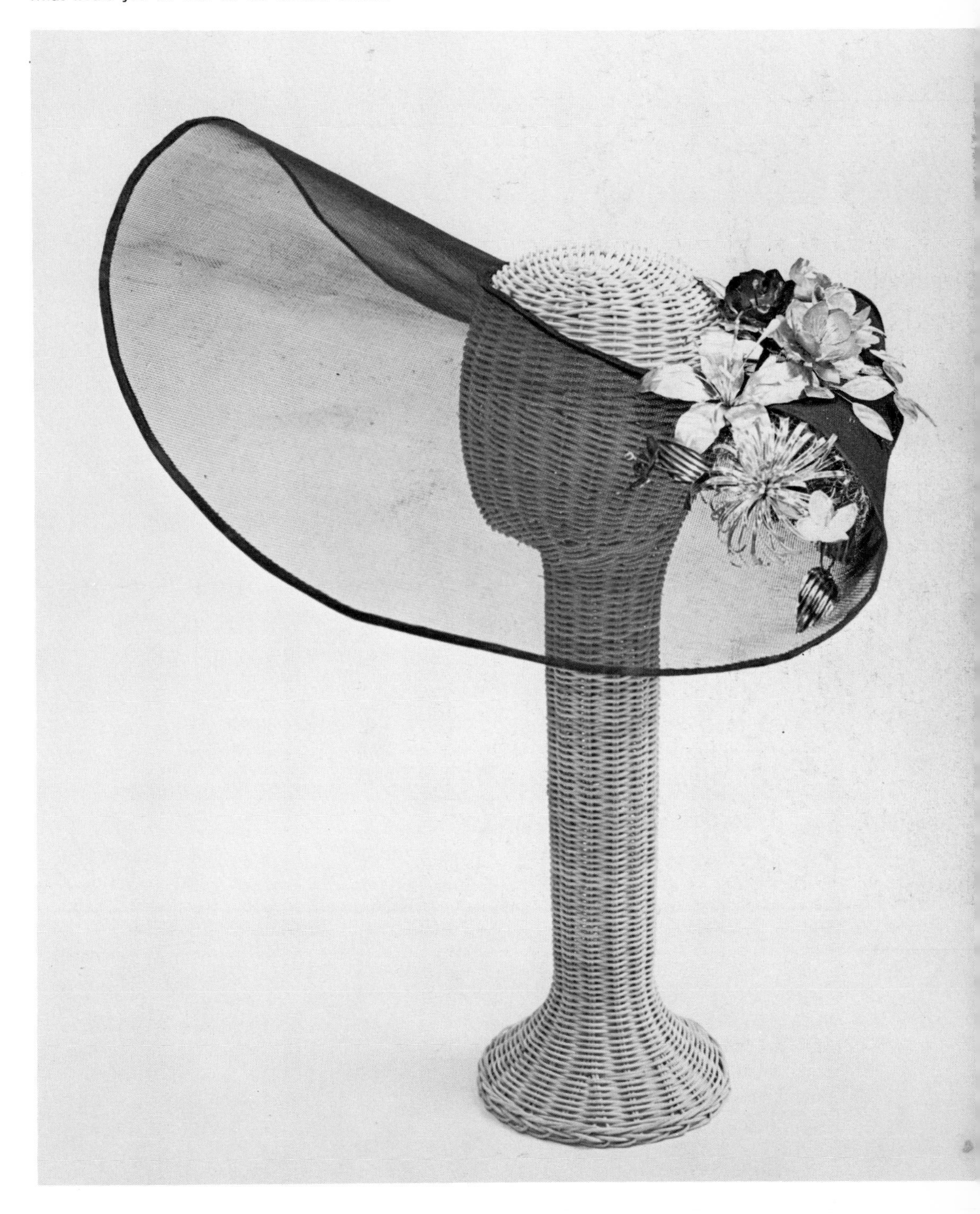

Nosegay for a Tenth Wedding Anniversary

Because the Tenth is Tin.

Not so long ago, the Museum of American Folk Art presented an engaging exhibition called *The Tinker and His Dam,* in which there were more than three hundred items gathered together on loan from public and private collections. The most amusing pieces were those designed for the tenth wedding anniversary. Such whimsies. Top hats, bonnets, dancing slippers, aprons, combs, fans, nosegays, even a tin feather boa!

TOOLS

Coarsely-serrated snips
Ball-pein hammer and board
Long-nosed pliers
Round-nosed pliers
Compass
Propane torch
Asbestos mat
Vice, clamp, screwdriver

MATERIALS

Large, round lid at least 8″ in diameter, or piece of tin from side of can 8″ square
Lid from 2-pound coffee can
6 lids from soup cans
5 lids from small frozen juice cans
16-gauge wire
Florist wire
Sash cord wire
Grease pencil or felt-tip pen, ruler
Solder, Dunton's Tinner's Fluid
Steel wool

THE IMPERIAL DOILY

So-called because the lace-edge is made with the Imperial Necklace technique (see page 174). Those of you who have made the Imperial Fireplace Fan will have a head start on this project.

Cut down the lid of a round commercial can to a circle 8″ in diameter, or cut one from the side of an unribbed can.

Inscribe another circle on it, 1½″ smaller, with a compass.

With ruler and grease pencil (or felt-tip pen), divide the circle into sixteen equal sections (see Dividing Lids Into Equal Sections, page 15).

Place a dime in the center of the circle and trace around it.

Cut away *half* of one of the sections *and* the dime-sized circle in the center (see diagram). Hammer until perfectly smooth.

Cut narrow wedges out of the perimeter according to the diagram to make true square sections for fringing. When you get around to the half-section, trim away the top, leaving the lower portion as an overlap.

Tin both faces of the overlap (see Soldering, page 20).

Shape the cone by pinching the little hole in the center with your fingers to help it curve and by pulling the cone around past the overlap. If that little hole curves smoothly, the rest of the cone will also.

To solder the overlap, clamp it firmly at the top.

IMPERIAL DOILY DIAGRAM

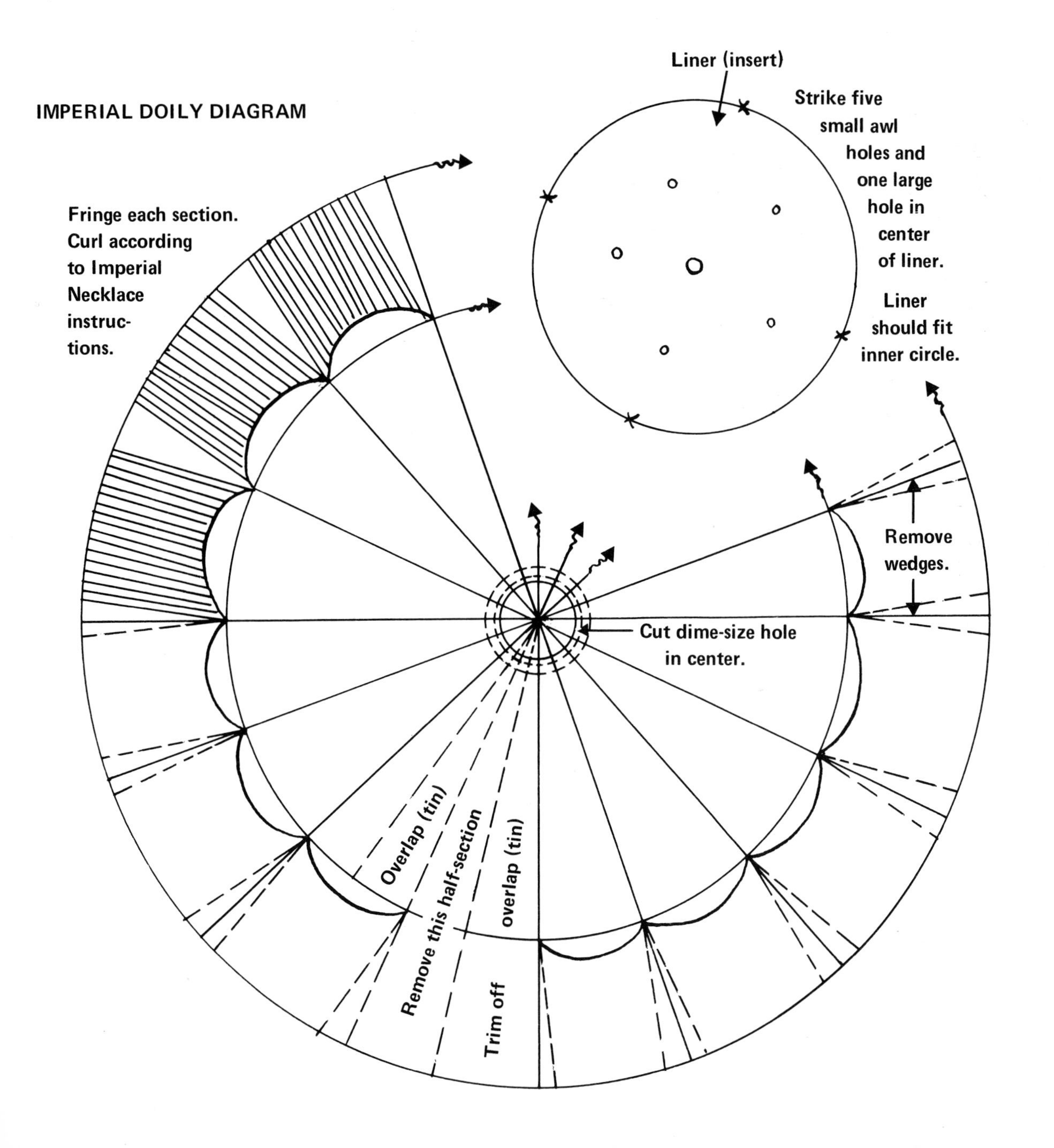

If the seam doesn't lie quite flat, pinion it with the screwdriver. Apply heat along the overlap until you can feel the solder flow. Remove the heat, but keep pressure on the screwdriver until the metal cools. Rinse with water and burnish with steel wool.

Before you give the doily the Imperial treatment, add the liner. This is made from the lid of a 2-pound coffee can in which you must strike five small awl holes for the filler flower stems to pass through and one large awl hole for the rose stems to pass through. (See liner diagram.)

When that is done, hammer the lid from the back with the flat of the hammer until it forms a smooth shallow saucer.

Trim ⅛" off around the edge of the liner and solder it, *convex side up,* in four places to the doily.

Give the doily its lace-edge, first drawing in the arcs freehand with grease pencil. Cut and curl the fringe down to the arcs with the coarsely-serrated snips, according to the Imperial Necklace technique on page 174, working every other section from the front, and the alternates from the back.

THE FLOWERS

Make your own selection of the flower forms in your bouquet. Most tight little bouquets or tussie-mussies use a lot of tiny flowers surrounding a red, red rose, but that's an awful lot of work. I took the easy way and put three Apothecary Roses in the center (that's why the larger hole in the liner there) and five small Plumbago blossoms around them. But whatever your choice, solder the leaves on those smaller flowers *close to* the blossoms (within an inch), so they can be pulled down snug against the liner and be framed by the lace-edge.

Solder the small flowers in position first—just a drop where their stems pass through the holes in the liner. Then insert the center grouping and solder them where the stems *emerge* from the cone *below.*

BIND THE STEMS

Because I somehow didn't like the look of all those stems going every which way at the base of the cone, especially since some were thick sash wire and some thin brass wire, I cut a long strip of tin, ¼" wide, and wrapped it round and round the full length of the stems and secured it with a drop of solder, top and bottom.

It might be nice to add a velvet bow and streamers. What do you think?

BAUBLES AND BANGLES

Having the sparkle of silver and gleam of gold, tin cans are everyman's treasure when converted to jewelry. Costing nothing but time, lacquered tin can be transformed to resemble the fabulous confections of forgotten civilizations when the world was virgin and wealth unmeasured.

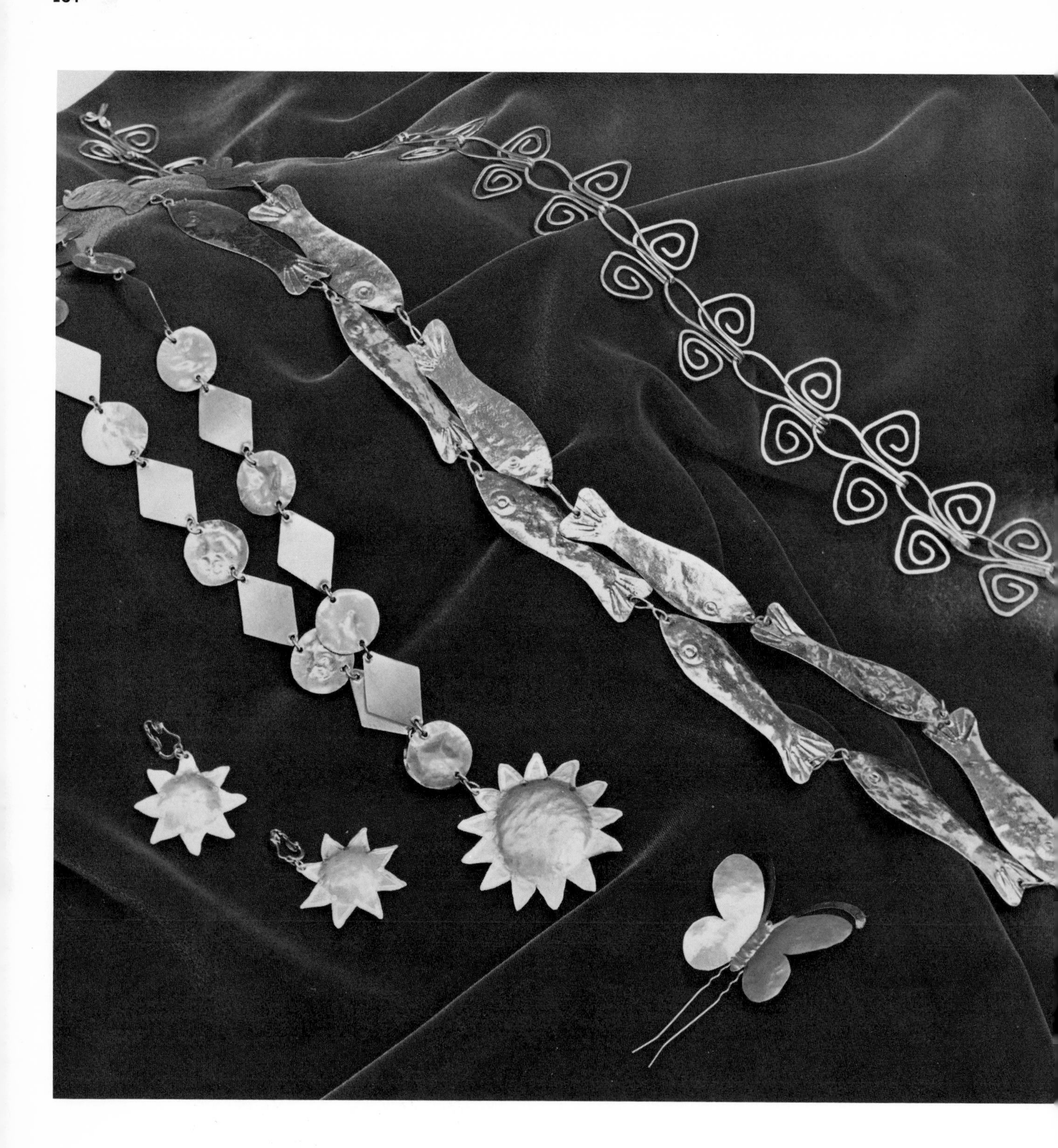

SUNBURST CHAIN AND EARRINGS

Effective, don't you think, for something easy to make?

TOOLS

Plain-edged snips
Ball-pein hammer and board
Awl
Long-nosed pliers
Kitchen chopping bowl
Dapping block

MATERIALS

Flattened sides of 2 soup cans, one gold-lacquered
3 matching gold-lacquered lids
Nu-Gold or florist wire, #1 knitting needle
1 pair earring findings
Giles' Black Varnish or jet-dry black enamel *(optional)*
Silver bronzing powder *(optional)*
#0 Fashion Design brush *(optional)*
Krylon Crystal Clear Acrylic Spray *(optional)*

By now you may have discovered my predilection for Campbell's Soup cans. See if you can't find one that is faintly lithographed in diamonds on the outer, silver side, and another (with a gold lining) lithographed in squares into which a quarter fits perfectly. Occasionally they come that way and make cutting out that much simpler.

But, lacking that luck, cut seventeen silver diamonds and eighteen gold suns from the flattened sides of two cans (Fig. 1). (If making a belt instead, you would probably want the pieces twice as big.)

Hammer the diamonds on both sides until they lie perfectly flat and have smooth edges. Round the points until they feel comfortable. Hammer them again. Trim and hammer the suns also.

FIG. 1 ACTUAL SIZE

Make 17 diamonds; 18 suns.

With hammer and awl, strike the awl holes, first from the front, then the back, hammering flat from the back.

Beat the suns on the silver side with the round end of the ball-pein hammer in the kitchen chopping bowl. Then, to make them look molten, place them, gold side up, on the hammering board and tap them lightly with the flat of the hammer.

Link the pieces together, alternating silver diamonds with golden suns, using either purchased links or those you've made yourself by wrapping wire around a knitting needle and cutting through the coil.

To make the sunburst medallion, divide the lid of a soup can into twelve equal sections (see Dividing Lids Into Equal Sections, page 15), first dividing the lid into quarters and those quarters into thirds. Cut slightly irregular flame tips in each section, rounding the tips for comfort's sake, and taking great care not to cut into the sun-center. You will lose the integrity of the design and that sense of an amorphous mass if your snips penetrate the center circle.

Having cut all the flames most carefully, hammer the piece flat, gold side down, working over the edges until soft and smooth. Place the piece, again gold side down, over that section of the dapping block which most nearly fits the area you want to raise, and beat it gently until it conforms to the block.

FIG. 2

Divide lid into quarters; then each quarter into thirds, cutting exactly to the center circle. Careful at these points.

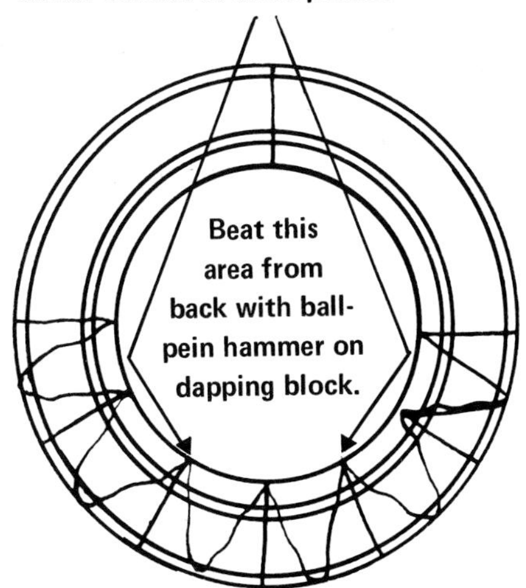

The earrings are done in exactly the same way. Make them a little smaller by cutting down the lid. Use *two* links in suspending them from the findings so they face forward.

Bronzing the medallion will enhance your careful crafting (see Pointers on Paints and Painting, page 17). Dip the blunt end of your #0 Fashion Design brush (or small brad head) into Giles' Black Varnish and make dots at those points where the flames meet (Fig. 3). With the bristle-end, paint in the fine border lines, being careful not to connect with the dots. Allow to become tacky and gently rub on the silver bronzing powder. Let dry overnight. Wash with soap and water. Apply protective coat of clear acrylic.

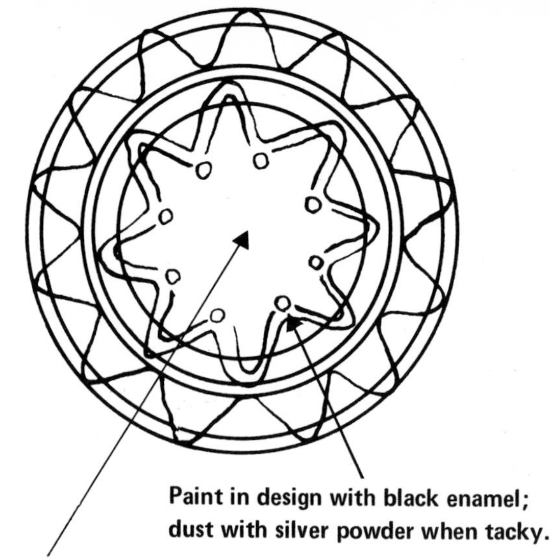

FIG. 3

SILVERY MINNOW BELT

Speaking of links, why not fishes? Circle your waist with a school of silvery minnows (Fig. 3). You can even cut them out with kitchen shears if necessary; beat them with the round end of the ball-pein hammer on a flat board to make them sinuous; strike in the tail lines with a screwdriver or chisel; glue on a rhinestone eye (or even easier, tap it in with a nail-set), and there you are!

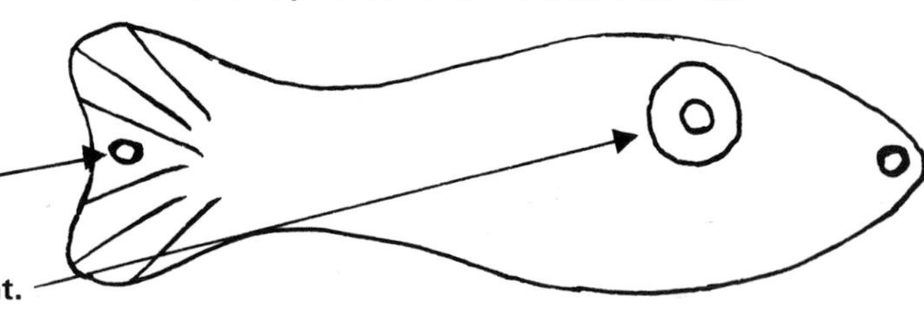

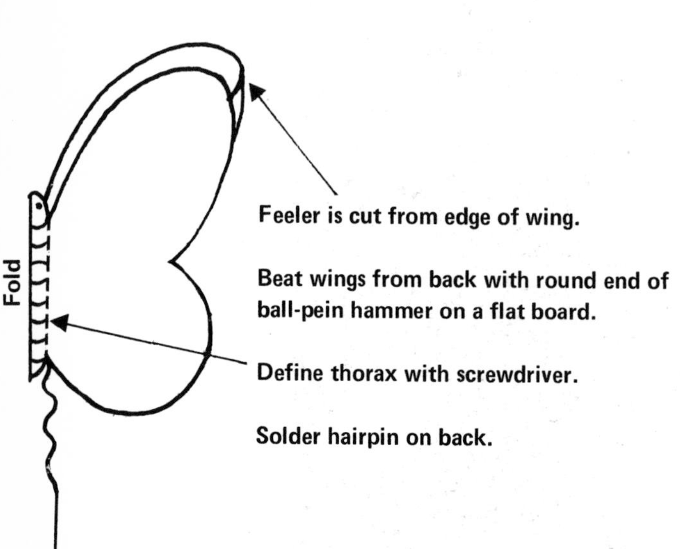

BUTTERFLY HAIR ORNAMENT

A butterfly, a bowknot, a single rose—almost anything simple and smooth in outline is suitable as a hair ornament. Simply cut out the design, embellish as you wish with punching, piercing or hammering, and solder a hairpin on the back.

The Royal Queen of Ur wore a comb of wildflowers on her padded wig—charming. And pretty for a party.

FLORENTINE CHAIN

This design is almost as old as civilization itself, and has been revived many times. It is currently enjoying such vogue that I thought if you already had the Nu-Gold wire, you might like to make it. Or order silver wire, if you prefer. (See Sources of Supply, page 195.)

For a belt you will need about eighteen 12″ lengths of wire. For a bracelet of smaller links, you would need about eight 7″ lengths. Make both coils, working with the round-nosed pliers and your fingers, *before* forming the loop.

You will notice that the loop, once formed, is folded under the coils and forward of them, toward the next link (Fig. 4). You will hook each successive link through the loop of the one before it. Notice the simple hook and eye clasp.

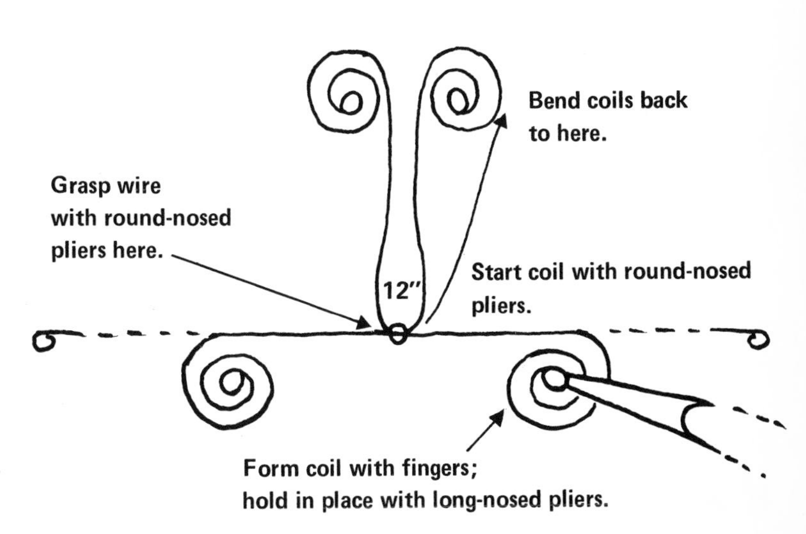

BANGLES

Obviously, the rim of a can is begging to be a bangle, but naturally it requires some disguising beyond being trimmed off, filed smooth, and burnished. You may want to flow on solder to make it look ancient. Or you may prefer to wind it about with Nu-Gold and florist wire and hammer it until "married." You will probably want to add a bauble of some sort, perhaps a bound coin or beach pebble.

All the designs you see here are crudely executed, but my teen-aged daughter assures me that that is part of their charm. Of course, you must suit your own style. Mine is inclined to be formal, but I do love that beach pebble, and find it quite formal enough when suspended from a circlet in the hollow of one's throat.

TOOLS
Snips
Round-nosed pliers
Ball-pein hammer and board
Fine file
Awl
Screwdriver
Penknife
Propane torch
Asbestos mat
Vise, short length of pipe

MATERIALS
Rims of cans
Foreign coins or tin discs
Nu-Gold wire
Florist wire
Earring findings
Steel wool
Solder, Dunton's Tinner's Fluid, baking soda
Carpet tack
Scotch tape
Weldit Cement or Epoxy 220

The rims from soup cans are about the right size around for the average hand, but there *are* rims just a shade larger if you need the extra room.

Test the rims for sheen by rubbing them with steel wool before you start trimming, because some will come up silvery-satin and others will grow increasingly steely, and there's no point in wasting time on them.

To make a rim presentable, trim off the raw edge with plain-edged snips. File any remaining unevenness until you get right down to the solid rim. Burnish bright with steel wool.

If you like the crude, molten look of solder, flow it on (see Soldering, page 20).

If you prefer the elegance which a touch of gold contributes, bind the rim with Nu-Gold wire, forming a loop as you finish it off if you wish to add a pendant. Follow around after the gold wire with ordinary florist wire. Slip the bracelet over a short length of pipe held in the vise, and hammer the wires *flat*. They will sink into the rim in such a way as to make them seem woven together.

COIN PENDANT

If you decide to bind a coin, choose one of foreign origin. (There's something unpleasant, maybe illegal, about defacing American currency.) Uncoil the spool of Nu-Gold wire carefully in order to avoid putting kinks in it and, with the round-nosed pliers, form a tiny loop on the end of the wire. Tack the loop onto your asbestos mat.

Place the coin on the pad directly under the loop and, holding it firmly in place, wrap the wire around it. Note precisely where the wire completes the circle under the loop and nip it off. Solder in place (see Soldering, page 20), holding the wire in position with a screwdriver.

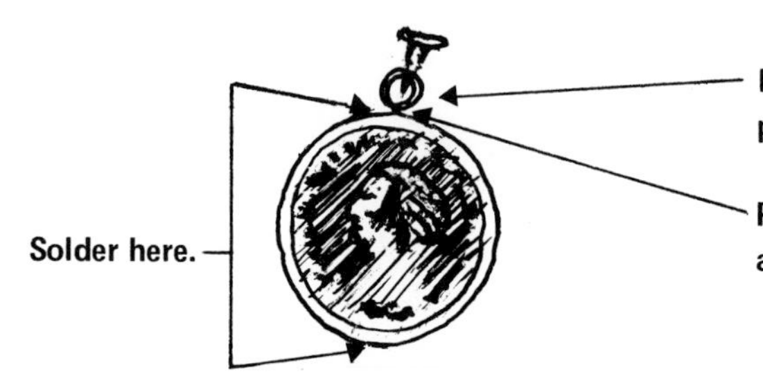

BEACH PEBBLE PENDANT

The procedure for binding a beach pebble is much the same, but requires a narrow strip of tin called a bezel in addition to the wire.

This strip should be no more than 1/16″ wide and be cut with great care and accuracy. Hammer it flat. Wrap it around the sides of the pebble, noting exactly where it overlaps. Make a true, vertical cut so the ends will butt perfectly (some trick!). Glue in place with Weldit Cement or Epoxy 220. Secure with Scotch tape to dry in warm oven.

In preparing the pebble for binding, remove any visible excess glue (with a penknife) from the two spots you expect to solder, because it will burn black and look unsightly.

With the round-nosed pliers, form a tiny loop on the end of the spool of Nu-Gold wire. Tack the loop to the asbestos mat and smooth the wire around the sides of the pebble. Nip it off where it meets under the loop.

To solder the wire onto the bezel, place the pebble, *with the loop end down, in the vise so that the wire rests in place on the bezel.* Carefully clean the spot to be soldered (see Soldering, page 20), slip a tiny strip of hammered solder between the wire and the bezel, and apply heat until the solder melts.

Place the pebble *upright in the vise and at right angles to the jaws* so that the wire is squeezed into position under the loop, and solder at the top. Clean with baking soda.

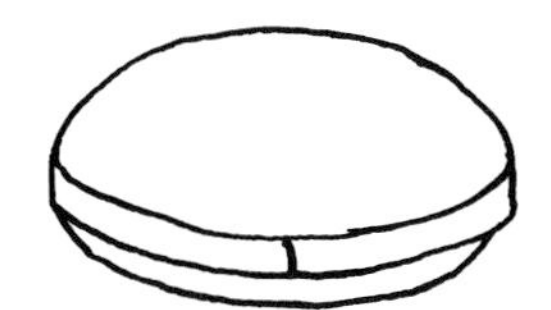

**Bind pebble with 1/16″ wide tin bezel.
Secure with glue.
Wind with wire, form loop for hanging and solder in place.**

PATTERN

Actual outline of 12-pound ham can lid.

Cut into a; bypass reverse curve; come back to it later; go on into b. Repeat at c and d.

Swoop around this U-curve.

Start cutting here, bypassing scallops.

SOLOMON ISLANDS CHOKER

If you can imagine it, the pendant on this choker was originally a Polynesian nose piece. No question, you and I would prefer to wear it over a turtleneck. A ham can lends itself perfectly to the design.

TOOLS

Plain-edged snips
Hammer and board
Kitchen chopping bowl

MATERIALS

Lid from a 12-pound Polish ham can
0000 steel wool
Nu-Gold or florist wire *(optional)*
Architect's Linen, scissors, pencil
Rubber cement

Trace the pattern onto a fold of Architect's Linen and glue the fold together with rubber cement before cutting it out. Open and apply to the hammered lid of a 12-pound ham can with rubber cement. Referring to the diagram, cut out the choker, noting open ends in back, working counterclockwise and bypassing all those little scallops at the bottom. Go back to them after you've done the rest.

Hammer the choker flat. Improve the outlines where you've had trouble. Hammer again until all edges are smooth. Burnish with steel wool.

Shape the pendant to make it slightly convex by hammering it from the back on a flat board. Shape the choker in the same fashion, tilting the sections at the back of the neck where it must stand up almost vertically in order to fit. The shape of your neck will determine the amount of curve you need.

You may prefer to *link* the choker to the pendant, rather than cutting the whole ornament from one piece. In that case, you would also have to connect the choker at the back of the neck with hook and eye.

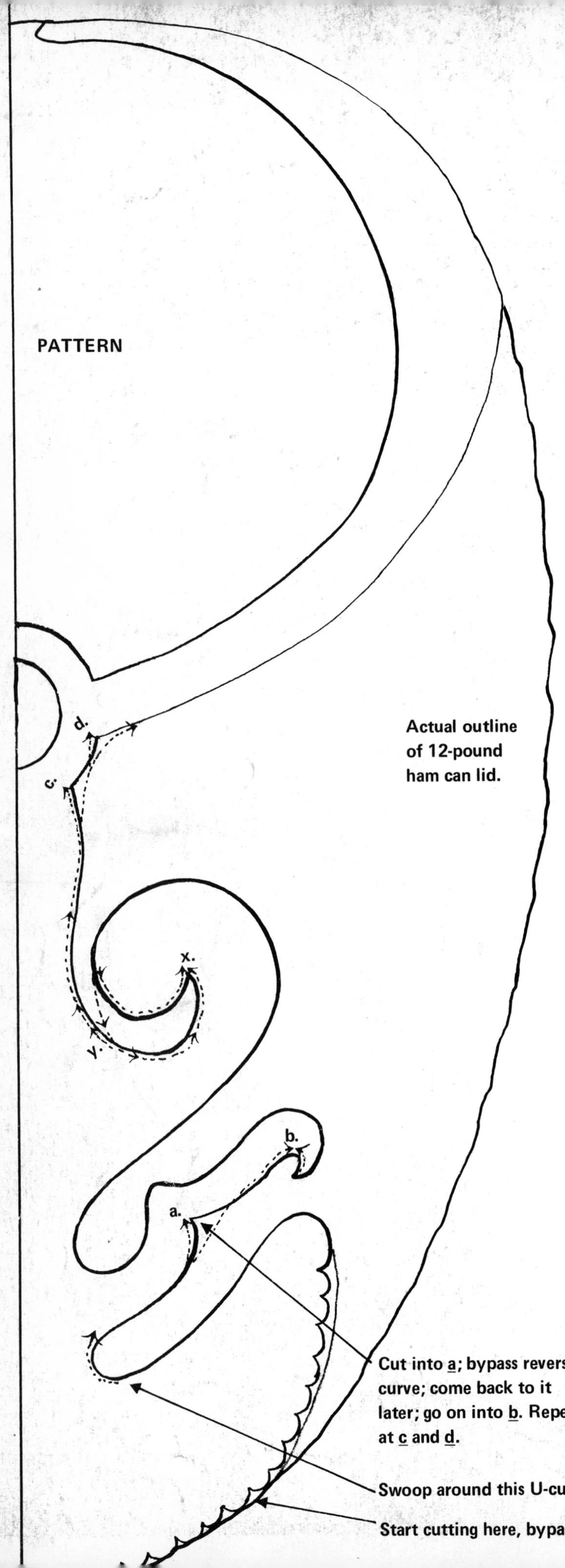

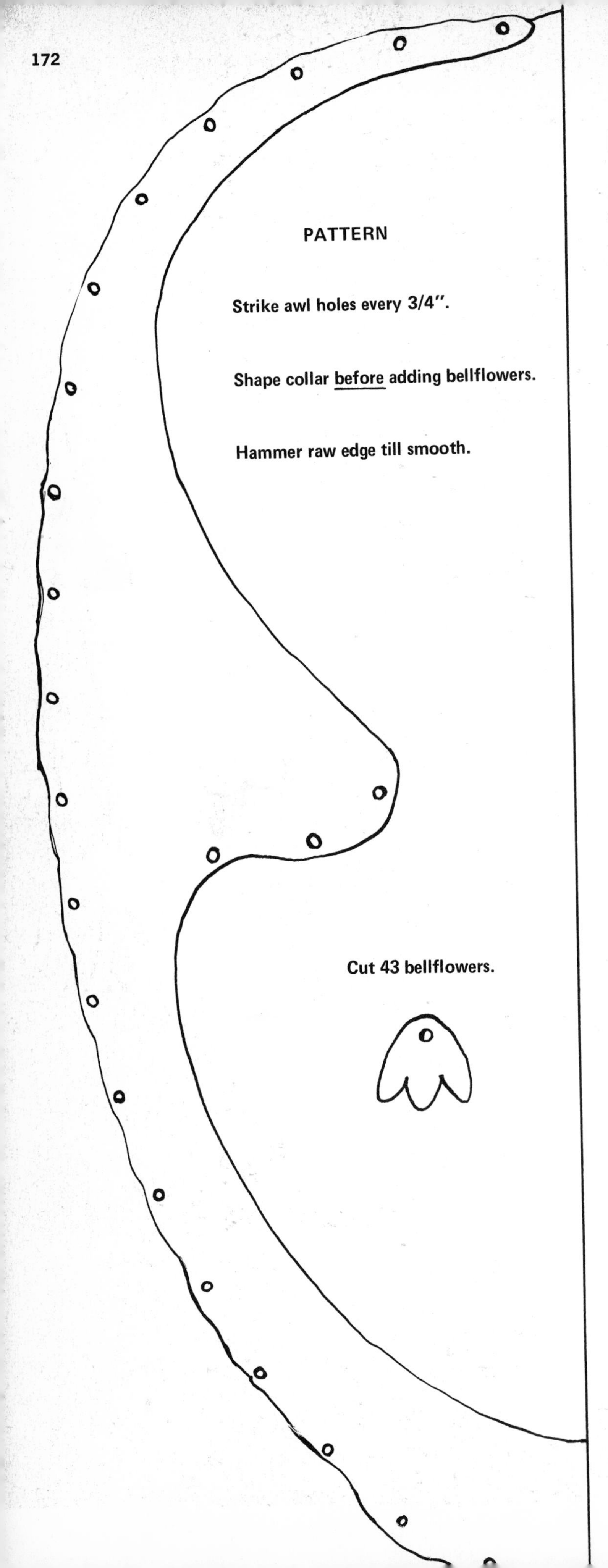

BELLFLOWER COLLAR

Reverse curves can be most intriguing. The design for this collar resulted from cutting out the vase shape needed for backing the French Urn of Flowers (see page 34). Admiring the bold simplicity of its outline, I could not bring myself to throw the remnant out, so here it is, complete with little bellflowers. They also were reverse curves, having come from all the U's cut out of the Staghorn Stars in *Tincraft for Christmas*. So, moral: save your scraps, especially if you have a lot the same shape, for eventually you'll find good use for them.

TOOLS
Plain-edged snips
Hammer and board
Awl
Long-nosed pliers

MATERIALS
Lid from 12-pound Polish ham can
Architect's Linen, scissors, pencil
Rubber cement
Nu-Gold or florist wire for links
Fine knitting needle

Trace the pattern on a fold of Architect's Linen and glue together before cutting out (see Tracing Patterns, page 16).

Apply pattern with rubber cement to the hammered lid of a 12-pound ham can and cut out, noting open ends on back. Hammer all edges until perfectly smooth.

Strike awl holes every ¾" around the outer edge, first from the front, then the back, hammering flat from the back.

Cut out 43 little bellflowers, fishes, or whatever, from the rest of the lid; nip any points, and hammer smooth.

Strike an awl hole in each and attach to the collar with links made by winding Nu-Gold or florist wire around a fine knitting needle and cutting through the coil.

IMPERIAL NECKLACE AND EARRINGS

Quite eye-catching and surprisingly easy to make, but you *must* have coarsely-serrated snips.

TOOLS

Coarsely-serrated snips
Hammer and board
Round-nosed pliers
Awl
Oval die or sawed-off screwdriver

MATERIALS

2 cans, unribbed, with bright gold linings
2 matching lids
Tracing paper, pencil, rubber cement
Grease pencil or felt-tip pen, turpentine
2 upholsterer's tacks or other center ornament for earrings
2 earring findings
Weldit Cement
Nu-Gold, or other wire, and fine knitting needle for links

Trace the pattern (Fig. 1) and apply with rubber cement to the flattened side of an unribbed can with a bright gold lining. Cut one blank; trace around it to make seven or eight more, depending on the size of your neck. Hammer them flat.

Improve the outline if necessary; hammer again.

Strike the awl holes for the connecting links (Fig. 2). (See Striking Awl Holes, page 14.)

Ornament the "wings" with die marks from front and back, or any way you wish (Figs. 2 and 3).

With grease pencil or felt-tip pen, mark off the triangles and cut the fringe, holding the third finger of your helping hand in the way as it curls down to keep it as straight as possible. Hammer the fringe flat, working from the base out to the tips.

Curling the fringe looks difficult because it is so fancy, but the actual procedure is simplicity itself.

Start by holding a blank, fringe up, in your helping hand. Curve the fringe over backward, one strand at a time, by pressing it between your thumb and fingers, pinching, smoothing, squeezing, so that the curve is soft and true.

Then, with your helping hand holding the base of each strand in turn, pull them out to right and left as shown in Fig. 4.

With the round-nosed pliers, put little hooks on the ends of the strands. Do this on the very tip of the strand with the very tip of the pliers, holding the strand with your helping hand right up close to the tip, so that only the tip turns, not the whole strand. If you get careless and don't work with just the tips and hold everything snug, you'll get a design so coarse and unrefined as to be unwearable.

The hooks are made with a 180° twist of the wrist. Working on the right side of the ornament, you will start with the back of your hand *up*, and twist your wrist over backward until your palm is up. On the left you will start palm up, and twist your wrist forward until the back of your hand is up. (All southpaws will reverse the procedure.)

The earrings are made according to the same principle, but are slightly more complicated because their sections overlap. Cut four wedges out of the lids, as indicated in Fig. 5. Fringe the four remaining sections and curl them.

Make links for the necklace and earrings by winding Nu-Gold wire around a fine knitting needle and cutting through the coil. Attach with pliers.

NECKLACE LINK PATTERN

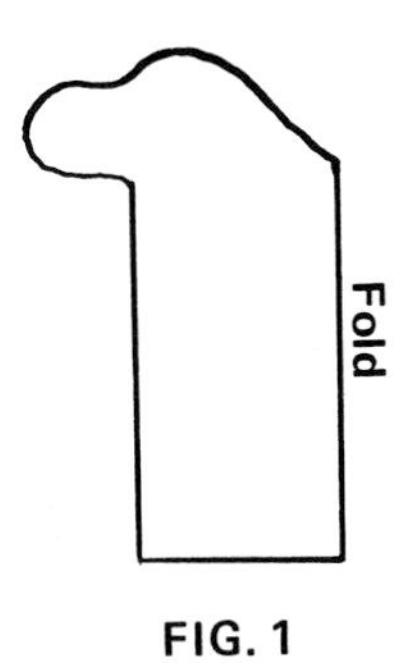

FIG. 1

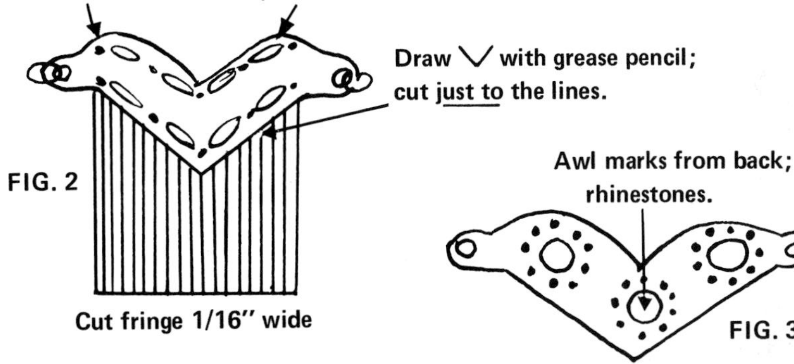

FIG. 2

FIG. 3

EARRING DIAGRAM

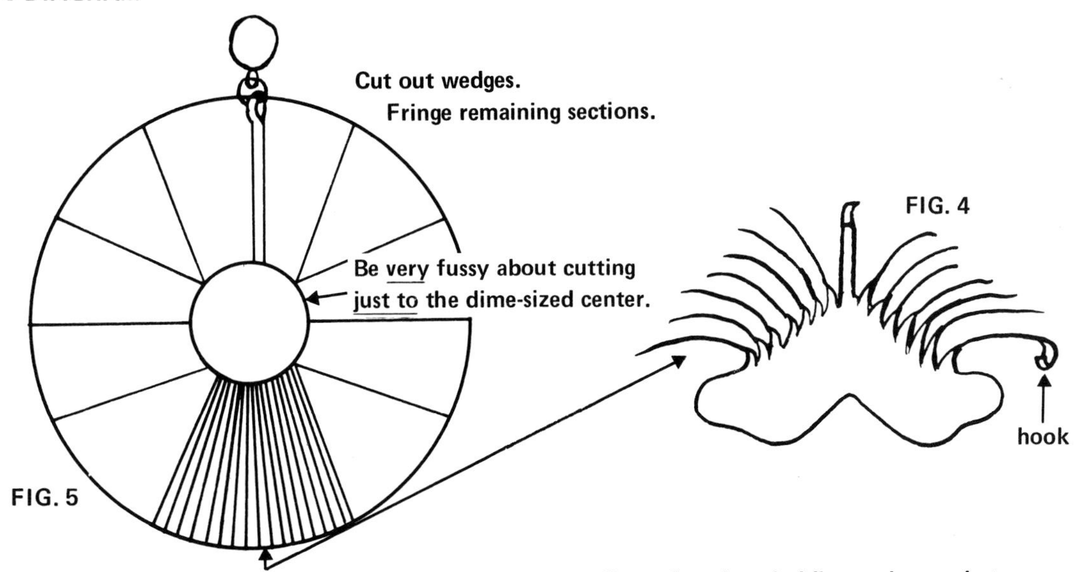

FIG. 4

FIG. 5

Curve each strand back, rubbing with thumb over fingertips; then, holding each strand at base, pull strands out horizontally to right and left. Make hooks with round-nosed pliers.

ASPEN LEAF NECKLACE AND EARRINGS

Consider Queen Shubad, mistress of the Royal House of Ur. She fancied the simple forms in nature—a leaf, a flower—attached with innocent artifice to her padded wig. She wore three leafy diadems like this at once, secured with a comb of wildflowers. Enormous loop earrings fell to her shoulders, and a single rose hung from the circlet about her neck. Today's big-name couturiers have taken their cue from Queen Shubad, and you can buy their copies of her jewels everywhere—Bergdorf's, Rive Gauche, Brentano's—but you'll like your own hand-crafted version better.

TOOLS	MATERIALS
Plain-edged snips	2 soup cans with rich gold linings
Hammer and board	40 tubular beads
Chisel	Gold cord
Round-nosed pliers	Nu-Gold wire or paper clip for catch
Awl	Tracing paper, pencil, scissors
	Rubber cement
	Weldit Cement
	Earring findings

There are lots of interesting, crude beads in the hobby shops these days, but you can make your own out of tin, or out of the Ronzoni pasta called *tubettini.* Roll the pasta in light and dark shades of acrylic paint (blue for lapis lazuli, red for carnelian, green for jade) and bake them on a cookie sheet for half an hour at 250 degrees.

Trace the leaf pattern and apply with rubber cement to the flattened side of a gloriously gold can. Cut out with plain-edged snips and use this leaf as a pattern for ten more, scribing around it with an awl.

Hammer the leaves flat. Improve their outlines, if necessary, and hammer again.

Vein them with chisel and hammer from the front as indicated on the pattern. Tap them lightly from the back to flatten them a bit.

Bend the stem-end of each leaf with the round-nosed pliers to form a loop.

Should you wish to make your own beads out of tin, cut several strips from the side of the second can. Hold a strip, gold side up, in your helping hand. Grasp the very end of it with the round-nosed pliers, and with a twist of the wrist outward and over, curl the tin around the pliers until an almost perfect tube is formed. Snip off the curl and squeeze it with the pliers to make its seams flush.

You can make a perfectly adequate fastening for the necklace with an ordinary paper clip (see diagram).

In stringing the leaves and beads, knot on one part of the clasp to a 15″ length of gold cord. (Christmas Tie-Tie is good if the beads have large holes.) Clip off the tag-end and daub the knot with Weldit Cement on a toothpick. Slip on the beads and leaves. Knot on the other half of the clasp, again clipping and daubing with Weldit Cement.

PATTERN

Cut eleven leaves for necklace; two more for earrings.

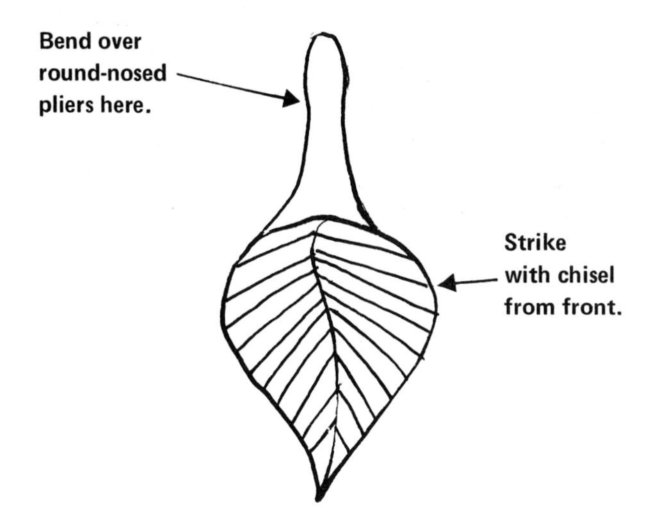

PAPER CLIP CLASP

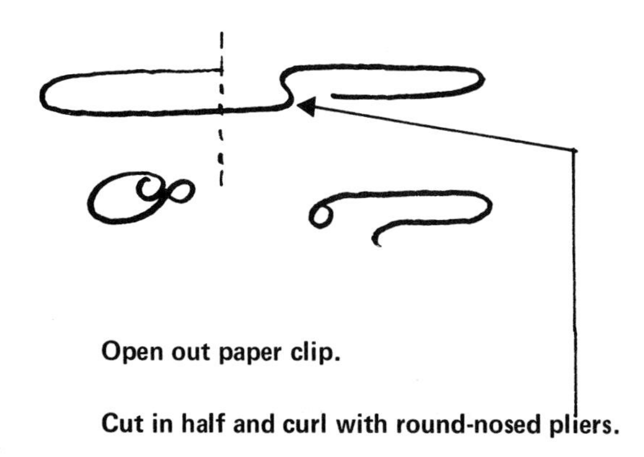

WILLOW LEAF NECKLACE

This dreamy confection is another of Queen Shubad's adornments. The original had three rows of beads suspending the leaves, and made a stunning collar, but the stringing was too complicated. However, just for fun, you might want to look it up in a fabulous book entitled *Jewelry Through the Ages* by Guido Gregorietti (American Heritage). Now *there's* a vol ume to impress you with man's artistry!

TOOLS

Plain-edged snips
Hammer and board
Round-nosed pliers
Small screwdriver
Filed-down nail

MATERIALS

5 soup cans with soft gold linings
Tubular beads resembling lapis lazuli, if possible
Architect's Linen, scissors, pencil
Rubber cement
Gold cord
Weldit Cement
Hair wire *(optional)*
Nu-Gold wire or paper clip for clasp

Trace the leaf pattern, carefully, onto the dull side of a piece of Architect's Linen, and apply with rubber cement to the flattened side of a can with a soft gold lining. Cut 15 leaves. Hammer them flat; improve the outlines, if necessary. Hammer again.

Define the central veins with small screwdriver and hammer.

Hammer in all the tiny veins, working from the tips of the leaf up toward the stem-end, switching to the filed-down nail as the design narrows. There's nothing difficult about executing these little veins; it's just tedious, that's all. And you'll get a kink in the back of your neck if you do too much at one time, so give yourself a break.

As you vein the leaf, it will curl up; so, turn it over and hammer it flat frequently: it won't destroy the design.

When the veining is finished, grasp the leaf at the neck with the round-nosed pliers; bend the "fleur-de-lis" top, first toward you, and then back over the pliers, to form a loop. Glue the fleur-de-lis in place on the back of the leaf and secure with a paper clip until dry. (Do this so neatly that you can wear the necklace silver side out.)

You may not want lapis lazuli beads if you never wear blue, but that's what the Sumerian Royal Crown Jeweler used. Perhaps some of you living near Indian

reservations can find something similar, but I had to resort to making my own out of pasta (see Aspen Leaf Necklace, preceding).

If you do use pasta, you will need a fine gold cord to pass through the narrow holes when stringing the necklace, rather than the heavier Christmas Tie-Tie. In any case, fashion a clasp with the round-nosed pliers and Nu-Gold wire (or a paper clip, as in the Aspen Leaf Necklace) and string the necklace, securing the knots with Weldit Cement or by winding with hair wire in the East Indian fashion.

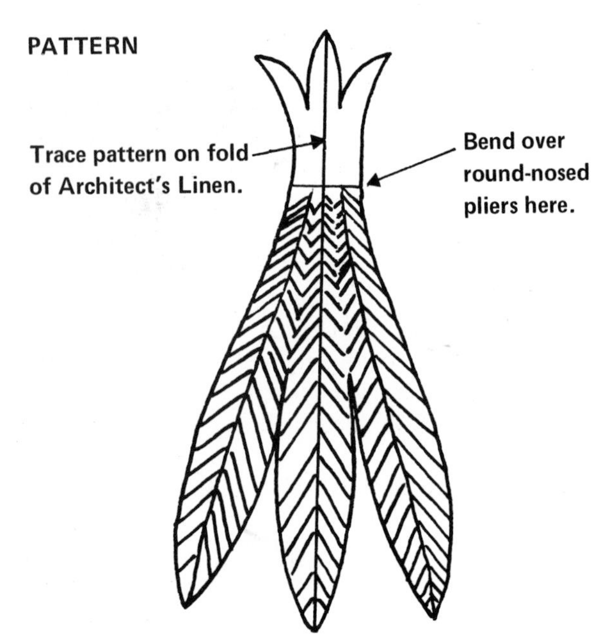

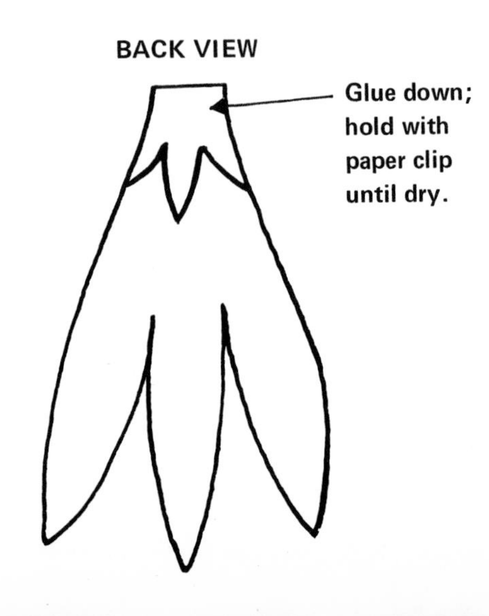

PART SIX
Christmas is coming!

CELESTIAL SPHERES A Study

Exercises in disciplined imagination, these starry wonders were Margery Strout's first creations in tin, incredible though that may seem. They reveal a native talent, annealed by an education at the Rhode Island School of Design.

STAR OF BETHLEHEM

If asked to choose one design which epitomized the potential of tincrafting, I would single out the Star of Bethlehem, together with its variations. For here a tin can has been transformed from "solid waste" into a symbol of light and life. See how easily, because of its classic simplicity and subtle coloration, it unifies the primitive Mexican *bulto* and elegant Adam mirror in Color Plate on page 136.

TOOLS
Coarsely-serrated snips
Hammer and board
Long-nosed pliers

MATERIALS
Small frozen juice can
Lid of Metrecal can
Paper and rubber cement
Krylon Metal Primer Spray Paint
Krylon Bright Gold Spray Paint
Small gold sequin
Sew-on-type rhinestone
Weldit Cement

Remove both lids and one rim of the small frozen juice can. Ornament the rim (see Ornamenting Rims, page 15).

Divide the sides of the can into *nine* equal sections (see diagram and Dividing Sides of Cans Into Equal Sections, page 15), making sure to center one end of the creased paper on the seam.

Cut along the creases in the paper, peel it off, and flatten the can out gently. You will notice that the rays tend to buckle near the rim, so hammer them carefully from the inside until you have persuaded them to lie out straight.

Spray-paint the lithographed side with three light coats of Krylon Metal Primer. When dry, spray-paint the other side with Krylon Bright Gold, unless it already is gold. Frequently it is not. At the same time, spray-paint the ornamented rim with Krylon Bright Gold.

Cut each ray as indicated in the diagram, holding the fingers of your helping hand in the way of the strands as they curve down to keep them from curling up tightly.

Hammer the shaft of each ray, starting at the base, to make it lie flat and even.

Then, grasping the tips of the strands, curl them slightly around your fingers in the direction they seem to want to go, spiraling them a bit to avoid angles, and encouraging the pairs to face each other (see photograph). If you get an awkward angle in a strand, straighten it with the long-nosed pliers. Notice that you leave the main stem of each ray uncurled.

Glue the gilded rim in place with Weldit Cement.

Make a small Frosty Star (diagram on page 120) from a Metrecal lid, if you have one. It's stiffer than other small lids and makes a crisper star.

Spray-paint the Frosty Star and the tip of each ray with Krylon Bright Gold.

Glue a small gold sequin and sparkling rhinestone in the center of the Frosty Star, and then glue it to the *back* of the main star, where it can twinkle secretly.

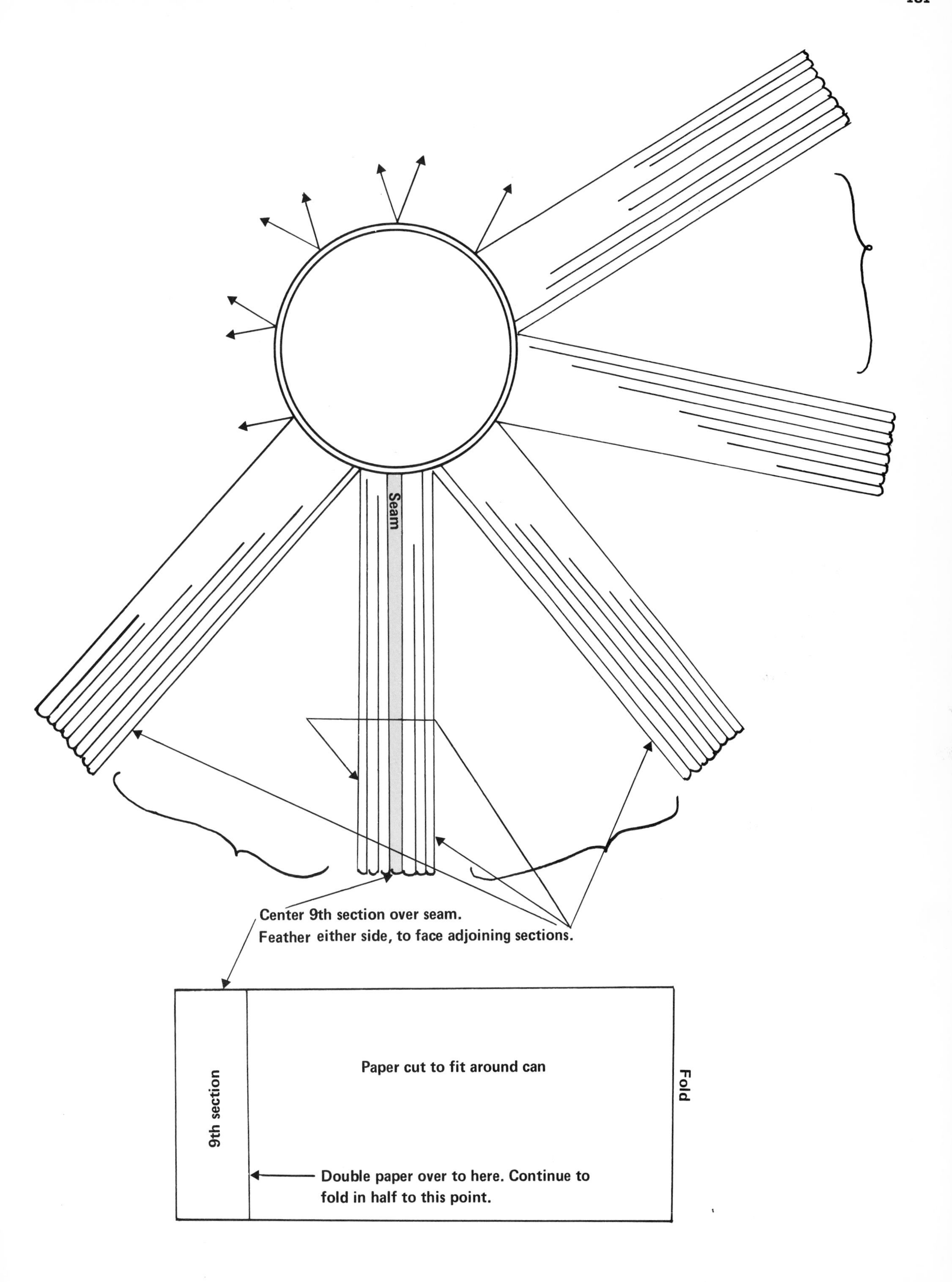
Seam
Center 9th section over seam.
Feather either side, to face adjoining sections.
9th section
Paper cut to fit around can
Fold
Double paper over to here. Continue to fold in half to this point.

EVENING STAR

A glorified version of the Star of Bethlehem, measuring eighteen inches across, the Evening Star makes a most effective wall sculpture. Refined and pure in form, it has a timeless quality about it which lends itself to both traditional and contemporary settings. Below you can see it blazing on my balcony.

TOOLS

Coarsely-serrated snips
Hammer and board
Long-nosed pliers

MATERIALS

V-8 Juice can
Rim from another V-8 Juice can
Lid from 3-pound coffee can
0000 steel wool
Fringed circle
Sew-on-type rhinestone
Weldit Cement
Krylon Bright Gold Spray Paint
Krylon Crystal Clear Acrylic Spray

Remove the bottom lid of the V-8 Juice can and divide the sides into sixteen equal sections (see Dividing Sides of Cans Into Equal Sections, page 15), making sure to lay one end of the creased paper exactly *on* the seam. Bend the rays out to the side; hammer flat; burnish with *0000* steel wool.

Cut each ray as indicated (see diagram), inhibiting

the curls as much as possible by holding the fingers of your helping hand in the way.

To make the strands lie flat along the shaft, run the flat end of the hammer *up into* the curls until they are pretty much back in place. Hammer the shafts flat from both sides.

Spiral the tip of each individual strand over your fingers in the direction it seems inclined to go, encouraging the paired sections to face each other. If you really fuss with them, you can make them as architecturally flawless as an Adam fanlight (see Evening Star Mirror, page 122).

Ornament the extra rim (see Ornamenting Rims, page 15) and spray-paint it with Krylon Bright Gold Enamel. Glue it in place with Weldit Cement.

From the 3-pound coffee can lid, cut a splendid Frosty Star (as in the diagram on page 105) and mount it in the center with Weldit Cement.

Protect it for always with a coat of Krylon Crystal Clear Acrylic.

EVENING AND MORNING STARS

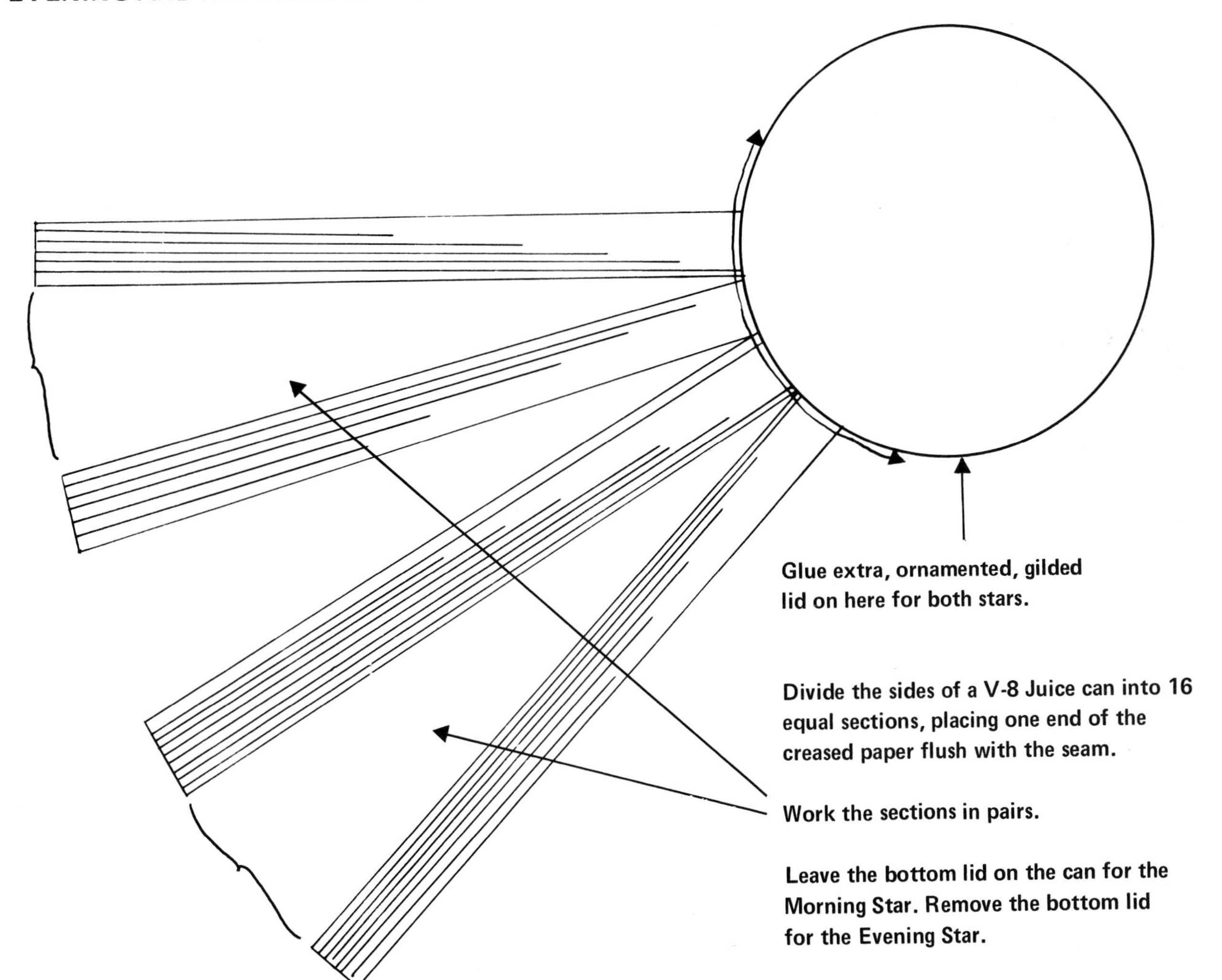

MORNING STAR

A variation of the Star of Bethlehem, the Morning Star is even easier to make because you simply hold the *third* finger of your helping hand in the way of the downward curving strands and *let them spiral off it.* You don't *do* anything about them; just hold your finger there and watch them glide off it.

Use the same technique on the star in the middle, which is made from the top lid of the can. Proceed just as you would for the Frosty Star (page 104), but hold that third finger in the way as you cut, and let the strands spiral off.

For those of you lucky enough to live near the source of supply in California, the center ornament is a eucalyptus seed pod, brightened by a rhinestone, but fortunately a V-8 Juice can is obtainable anywhere. Leave the end of the can intact and, gold side up, glue the center star upon it.

Also, if you like those "snigs" at the tip of each ray, leave the top rim on and cut right through it, borrowing a rim from another V-8 Juice can to ornament the center circle. Tools, materials, and general procedure are essentially the same as those in the Evening Star, preceding.

FESTIVE SWAG

A tin can is so perfectly suited for this purpose, it might have been made to order. And the tassel, too, mounted on a champagne cork which already has its own tin top, is a natural. (See color photograph on page 135.)

TOOLS

Snips
Hammer and board
Awl
Long-nosed pliers
Nail set and toothpaste cap

MATERIALS

Sides of finely ribbed cans
Lids from small frozen juice cans
Upholsterer's tacks
Weldit Cement
Gloves *(optional)*

THE ROSETTES

Cut down the lids from the small frozen juice cans to 1½″ in diameter. Hammer them flat. Strike an awl hole in the center of each, and dimple them with nail set and toothpaste cap from front and back, referring to the Federal Fanlight rosette diagram, page 129, for the simple how-to.

THE RIBBONS

You may want to wear gloves for this and read the section in Techniques on Opening Out Cans and Cutting Curved Ribs, page 14.

Remove the bottom lid and both rims from a can (preferably) with two sets of fine ribbing in its sides. Fruit usually comes in them.

Open up the can by cutting along the left side of the seam and pulling it open carefully, so as not to put a crease in the ribs.

Holding the can, concave side up, cut out the two sets of ribs (¾″ wide is about right).

Place them across each other to form an *X* and strike an awl hole through their common center.

If you plan to paint them, do so now before you fasten them together, unless you want the rosette painted too.

When dry, apply Weldit Cement between the layers, insert the stem of the upholsterer's tack through the rosette and the ribbons, and bend it over on the inside with long-nosed pliers and hammer. Secure with another drop or two of Weldit.

When the cement has hardened, *encourage* the curve in the ribbons so the paired tips almost meet. They will then grasp the garland securely.

The ti (pronounced tee and no relation to the ti leaves used in flower arranging) garland, incidentally, is the very best fake stuff I've come across. It's a good dark green (with only a little lightening at the tips), and has fine leaves rather like American box, altogether classic and restrained enough to satisfy the most discriminating taste.

THE TASSEL

What a whimsy! I just love it. I've always hated to throw away champagne corks, they have such character and cheery association, and finally I've found the perfect use for them. Simply fringe a piece of lacquered tin 3″ x 6″, inhibiting the curls by putting the third finger of your helping hand in the way as you cut (see Fringing Tin, page 15). Wrap the fringe around the cork and cinch it with a twist of florist wire. I soldered mine because it pleases me to do so, but it's not necessary. It will seem especially tassel-y if cut with coarsely-serrated snips.

Merry Christmas!

HEAVENLY BODIES

They say there are no two snowflakes alike, and I honestly believe that if everyone in the world were asked to make a star, each one would be different. Even working within the limitations of a single medium, the possibilities seem infinite. In addition to the variations on the Star of Bethlehem, there are four quite distinct forms in this group of Heavenly Bodies; and in *Tincraft for Christmas* there are many more; and they all "just happened." They required almost no conscious effort—they seemed somehow to be the natural expression of the tin.

Moreover, stars have almost as many uses as they have shapes. In the photograph, at the top left there is a treetop star; at the top right, a ceiling light canopy; at center left, a pin; on the lower left, a tree ornament (which could equally well be an earring); and on the lower right, a window ornament. As you can see from many of the color photographs in this book, stars can enhance plant tubs, chandeliers, and flower arrangements, let alone wall sculptures, fireplace swags, and barn doors. Like diamonds, they seem to be "forever."

This group of Heavenly Bodies is so easy to make that the photograph alone should suffice, but let me give you a few pointers. When making a treetop star or ceiling light canopy, crease the paper as

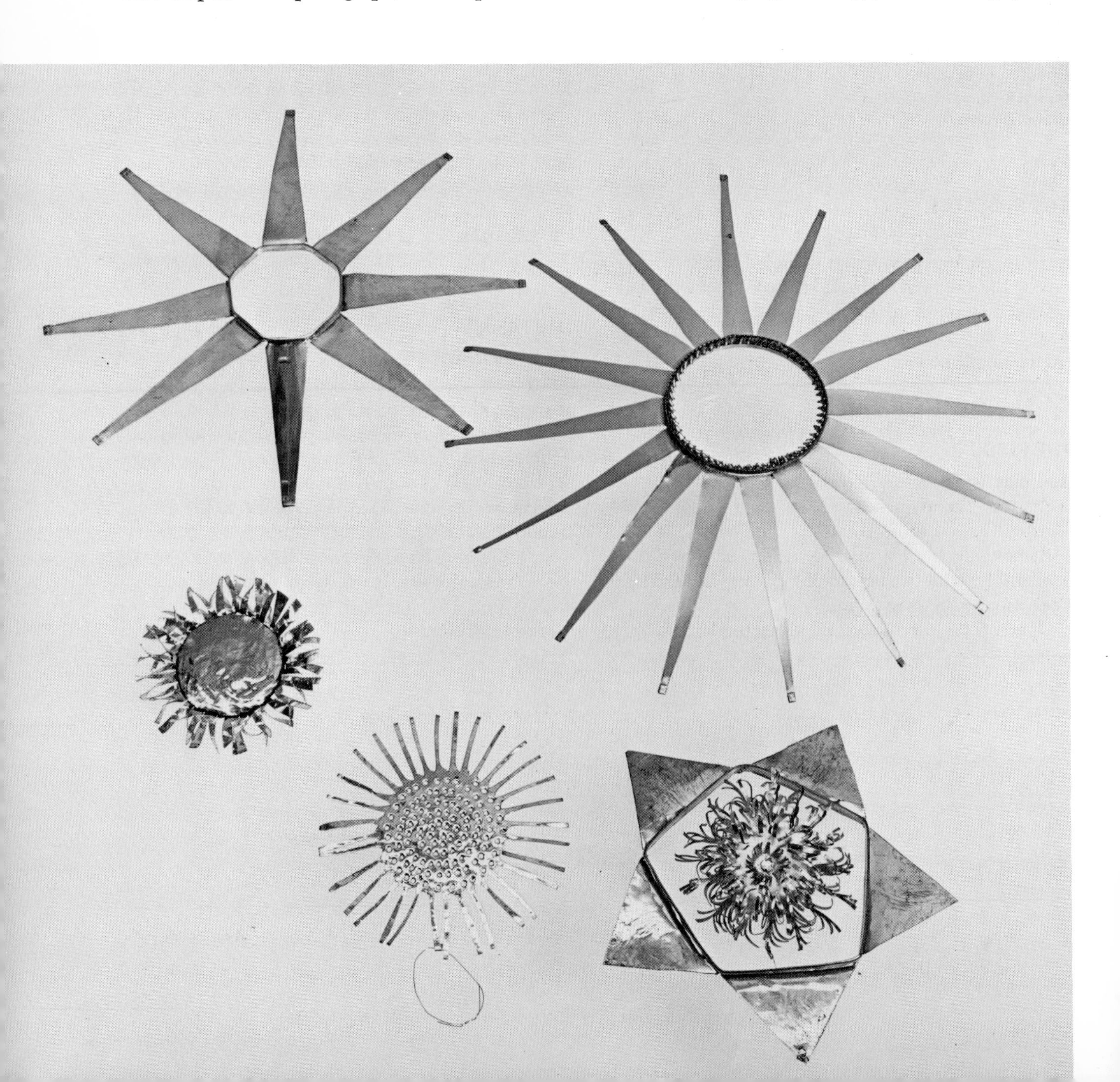

usual (see Dividing Sides of Cans Into Equal Sections, page 15), but *before* you glue it on, draw in the isosceles triangles in each section with ruler and pencil, and cut along *only those lines,* to save time and effort. Slip the end of the can over the corner of a counter to *flatten the rim at the base of each triangle.*

Similarly with the window ornament (lower right of photo), draw in the triangles on the creased paper before gluing it on the can. Incidentally, this *five*-sided star requires a little mathematics; you will have to measure the circumference of a can (perhaps cat food), divide by 5, and rule off the sections on the strip of paper cut to fit. As with the treetop star, hook the can over the corner of a counter and hammer the rim flat along the base of each triangle.

The pin (center left of photo) was the very first thing that Mrs. Joseph Wahl Roberts had ever made after getting her copy of *Tincraft for Christmas,* and she came up with something entirely original. I wish you could see it in color, it has such style: a rich gold center, ringed around with black enamel, mounted on a silver corona. Even though it is crudely cut, it is remarkably sophisticated and civilized, and the chasing marks, made with the *edge* of the *flat* end of the hammer, are particularly suggestive of an amorphous mass.

The pierced star (lower left of photo) is actually a reproduction of a form of plankton, or microscopic sea animal. I came across it in *Travel* magazine, which published an article on these fantastic creatures. I mention that only because, once you become aware of the potential of tin, you'll find your reading an inexhaustible source of ideas.

PAPER PATTERN

TREETOP STAR

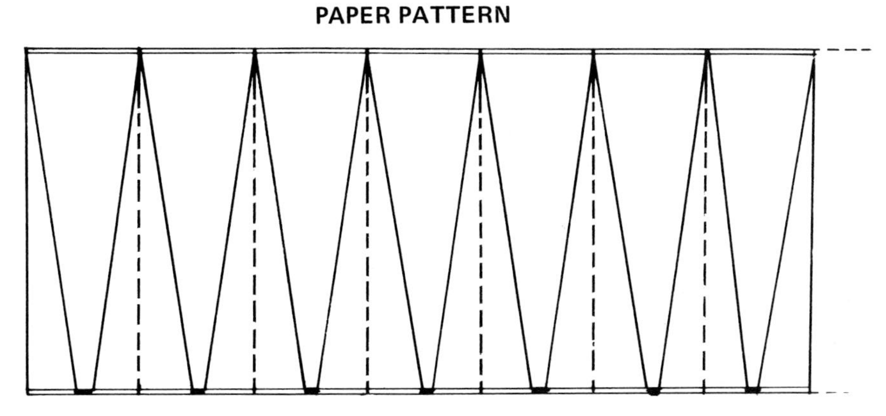

Cut paper to fit around can; fold in approx. 3/4-inch sections; and draw triangles in folded sections before applying paper to the side of can with rubber cement. Cut along solid lines only.

CEILING LIGHT CANOPY: Again, draw in triangles on folded paper. Coarsely-serrated snips would bead the edges prettily.

PLANKTON ORNAMENT: Plain-edged snips best here.

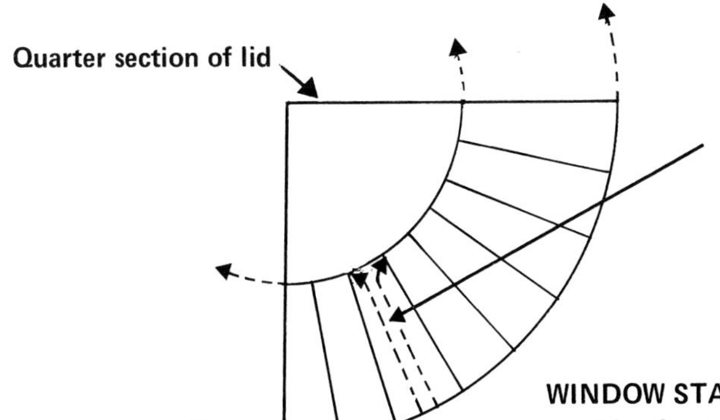

Hammer (1-pound coffee) lid flat and divide into 32 sections.
Trim away sides of each section, curving around to right and left at base of each section.
Stipple center portion with hammer and awl. Beat from back with round end of ball-pein hammer in kitchen chopping bowl.
See May Basket diagram for similar, simple star shape.

WINDOW STAR: Again, draw triangles on paper cut to fit before pasting on side of can. Frosty Star in center.

CHEERFUL CHERUB

Five cheerful cherubs greet you as you enter the Silvermine Tavern in Norwalk, Connecticut, where they light the entrance hall. In color photograph on page 135, see how ours beams down upon us throughout the Christmas season.

TOOLS

Plain-edged snips
Ball-pein hammer and board
Awl
Kitchen chopping bowl

MATERIALS

Sides of a 3-pound coffee can
2 lids from a 2-pound coffee can
Tracing paper or Architect's Linen, scissors, pencil
Rubber cement
Weldit Cement or Epoxy 220
Steel wool
Artists' oils and brush

Substitute an unlithographed can for the 3-pound coffee can, if you can find one (Idahoan potatoes come in them). Otherwise, work on the unlithographed side of the can, hammering it flat and burnishing it bright with steel wool.

Apply wing pattern with rubber cement and cut out. Hammer flat.

Beat one of the two lids in your kitchen chopping bowl with the round end of the ball-pein hammer, until it conforms to the bowl.

In the other lid, make a hole large enough to admit a nailhead. Glue the domed lid onto the flat lid, thus forming his head. Glue his head onto the wings.

Cut eye patches and apply to the lid with rubber cement. Paint the head and upper sides of the wings mustard. When dry, add features and wing lines.

PATTERN

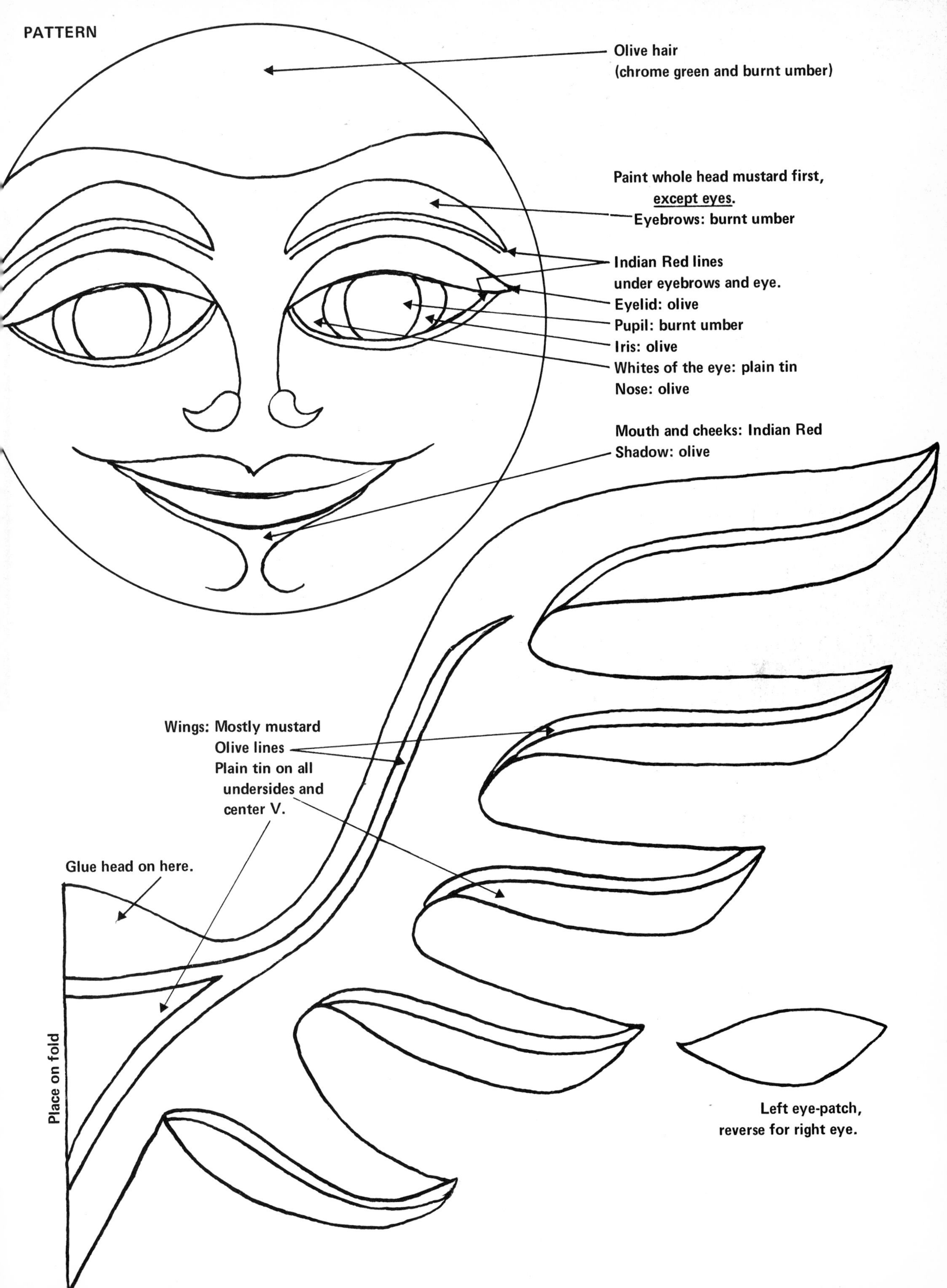
Olive hair
(chrome green and burnt umber)
Paint whole head mustard first,
except eyes.
Eyebrows: burnt umber
Indian Red lines
under eyebrows and eye.
Eyelid: olive
Pupil: burnt umber
Iris: olive
Whites of the eye: plain tin
Nose: olive
Mouth and cheeks: Indian Red
Shadow: olive
Wings: Mostly mustard
Olive lines
Plain tin on all
undersides and
center V.
Glue head on here.
Place on fold
Left eye-patch,
reverse for right eye.

PENOBSCOT INDIAN WIND CHIMES

In Old Town, Maine, the Penobscot Indians celebrate festive occasions by wearing these narrow bells sewn in clusters on their deerskin dresses and jerkins, making music with every motion and the atmosphere bright with sound. Depending on the size, the bells lend themselves to jewelry or wind chimes. Here the chime is hung from a scroll tacked to the top of the front door, where it swings and sings as people come and go—celestial at Christmas, but songful at all seasons.

TOOLS

Snips
Hammer and board
Awl
Round-nosed pliers
Long-nosed pliers

MATERIALS

Lid, and sides, of a ribbed can 6″ in diameter
2 lids from 2-pound coffee cans
2 lids from 1-pound coffee cans
Grease pencil *(optional)*
Florist wire
Small hank of rug yarn

THE HOOPS

From the sides of a large ribbed can about 6″ in diameter (dehydrated mashed potatoes, etcetera), cut three ribbed strips, ½″ wide, to make hoops from which to hang the bells (see Opening Out Cans and Cutting Curved Ribs, page 14). Make the strips 16½″, 12½″, and 6½″ long, respectively. Recurve them around a mixing bowl or other can.

To fasten the hoops, strike two awl holes in one end of each strip as shown in Fig. 1 (see Striking Awl Holes, page 14). Bring the other end of the strip around to overlap ½″, and strike corresponding holes in it *through* the first set. Wire the hoops together.

To suspend the hoops, stand them up on their sides as in Fig. 2, and make three equidistant holes near the edge of the rims with hammer and awl.

Cut three 7″ strands of yarn for the largest hoop, three 8″ strands for the middle hoop, and three 9″ strands for the smallest hoop. Knot *one* end of each strand. Working from the inside, pull the strands through the holes in their respective rims. Set aside while you make the bells.

THE BELLS

Before fashioning the bells, cut eighteen 9″ strands of yarn and knot *each* end. Later you will attach bells to *each* end of these strands and knot them over the hoops.

Cut a 15″ strand for the low center bell. Knot *one* end.

Cut two 3″ strands for the bells on the bow. Knot *one* end.

To make the bells themselves, divide the 6″ diameter lid from the can with the ribbed sides *and* each of the four coffee can lids into eight equal sections (see Dividing Lids Into Equal Sections, page 15). (This will give you 40 pie-shaped pieces altogether. Since you will only need *seven large* bells, experiment with the extra eighth piece until you get the hang of it; then proceed with the others.)

Cut off the point, as shown in Fig. 3. Hammer the pieces on a flat board to make them curve up (see Importance of Hammering, page 15).

Starting at the outer edge at the bottom of the bell, grasp the tin with the round-nosed pliers, palm toward you. With little *inward* twists of the wrist, work across the bottom of the bell, pinching the tin with the pliers and bending it *toward* you ever so slightly with each twist, as in Fig. 4. If you "inch" along in this fashion, moving along no farther than the width of the pliers with each twist, you will make a smooth, continuous curve so that later, when you squeeze the bell together with your fingers, the metal will be willing to curve around properly, rather than getting a crease in it.

FIG. 1

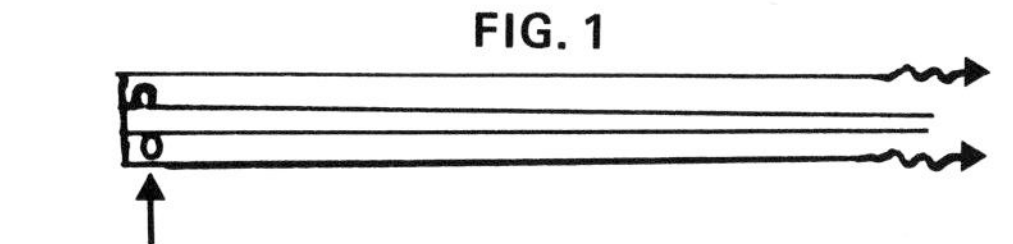

Cut 1/2″ wide strips for hoops from side of ribbed can; strike awl holes in ends and fasten with wire.

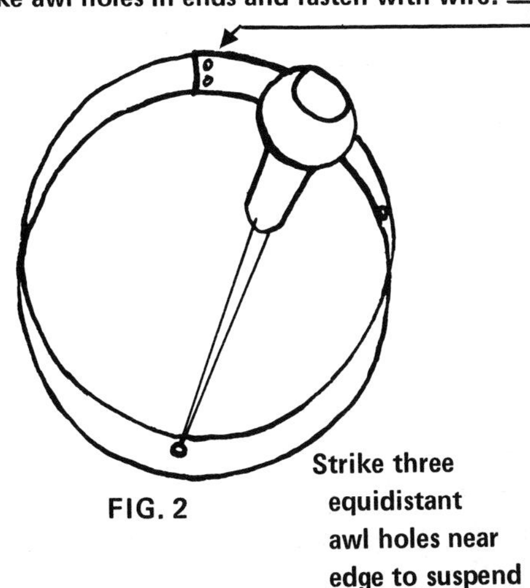

FIG. 2

Strike three equidistant awl holes near edge to suspend hoops.

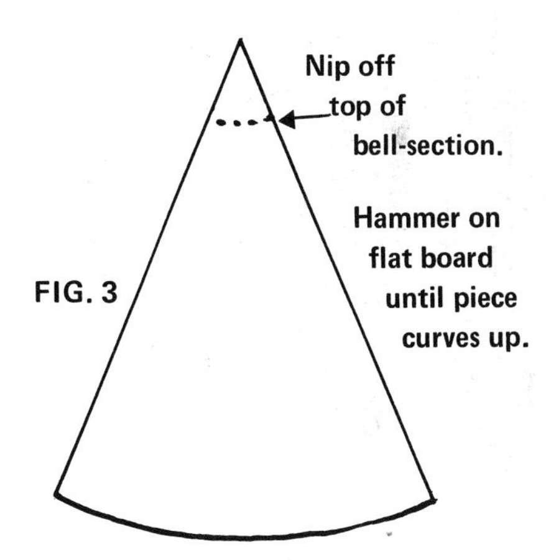

FIG. 3

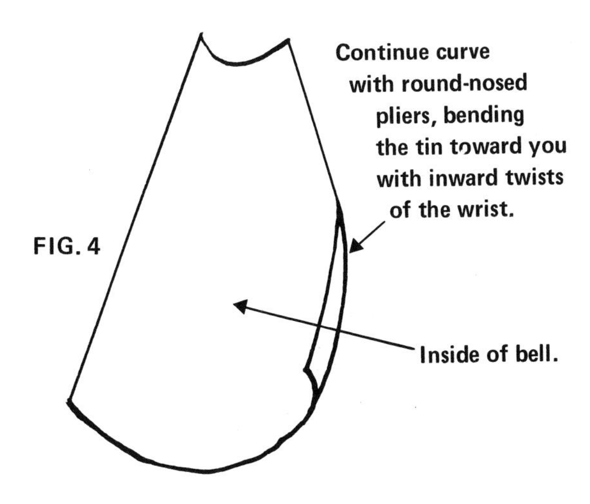

FIG. 4

When you have come as far toward you as you can, turn the piece over and work the opposite side of the bell away from you, as in Fig. 5.

Then come down from the top on the inside, and holding the pliers at an angle, as in Fig. 6, coax the out-of-shape edge into a nice straight line again, with little *inward* twists of the wrist, bracing your hand against your stomach. Similarly, straighten the opposite edge of the bell with little *outward* twists of the wrist. Fuss with it and tease it, until it resembles Fig. 7.

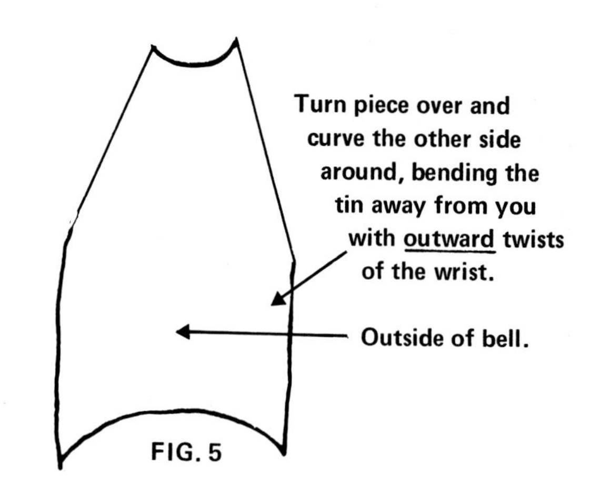

FIG. 5

Now grasp the bell firmly with the pliers from the top, and carefully pinch the lower section together part way *with your fingers.*

Then withdraw the pliers until only the tips are holding the bell, and pinch the top section of the bell together part way, again with your fingers, until it looks like Fig. 8.

Switch to the long-nosed pliers. Working down from the top, squeeze the bell just enough to flatten the right-hand edge, so that it can pass under the left-hand edge, as in Fig. 9.

Insert one of the knotted strands of yarn at the top of the bell. Holding the flattened edge of the bell with the pliers, as shown in Fig. 10, squeeze the bell together with your fingers, overlapping the seam a bit. Flatten the overlap by squeezing the bell gently from top to bottom with the long-nosed pliers.

[All this time you have been pinching and pushing with your helping hand—not to mention your stomach!—shaping the metal into a cone. Soon you will learn how each plier works and what it can do for you (see Effect of Other Tools Upon Tin, page 15). Basically, the round-nosed pliers *curve* the tin and the long-nosed *flatten* it. If the bell is not perfectly cylindrical, you can straighten the "bends" with the long-nosed pliers, always going back in the end to tighten the seam, which may have opened up.]

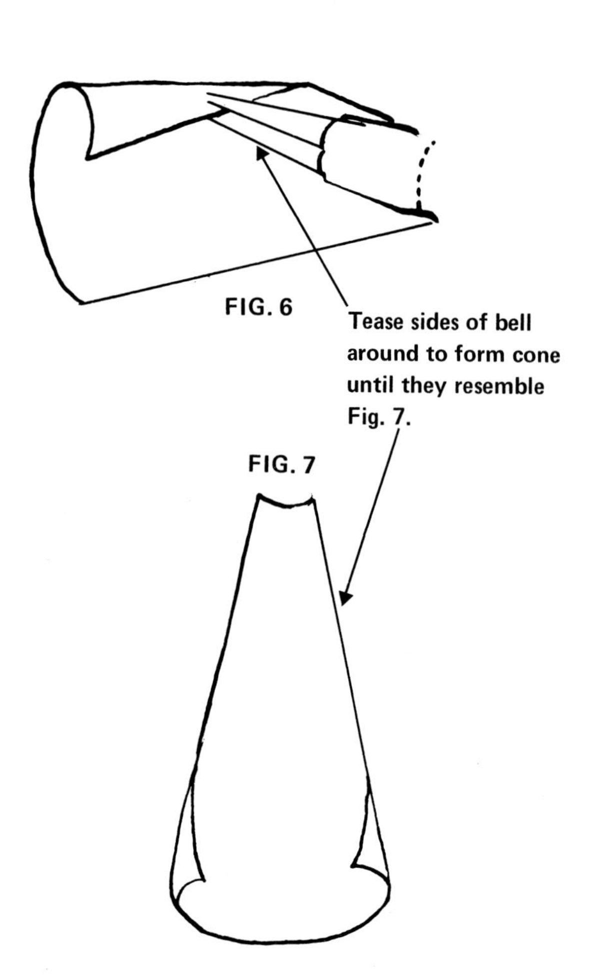

Altogether, for this chime, you will need sixteen small bells (made from the two 4″ lids); sixteen medium-sized bells (made from the 5″ lids); and seven large bells (made from the 6″ lid). That may seem like a lot of bells, but together they sound like a choir of angels, which will give you lasting pleasure. Once you get the knack of shaping them, you'll be surprised to see how fast they accumulate. Don't forget to insert the knotted strands as you go along. It's much easier than poking them through the top after they've been closed.

Attach one large bell to the 15″ strand and two large bells to the 3″ bow strands. Fringe these bells ½″ deep.

ASSEMBLE THE CHIME

Hang the strands with the sixteen smallest bells over the largest hoop, spacing them evenly. Knot them in place.

Hang six strands with medium-sized bells over the second hoop and knot them in place.

Finally, hang two strands of medium-sized bells and two strands of large bells on the lowest hoop, alternating the sizes, and knot them in place.

When it comes to putting the three tiers of bells together you will wish you had the four arms of Siva, but persevere! Pick up the strand of the low center bell and pass it up through the strands of the smallest hoop. Grasping those four strands together by their tips, pass them up through the strands of the middle hoop, etcetera. When you get everything pulled through, lay the chime down carefully and gather all the strand-tips together and bind them 1½″ down from the top with a 6″ length of florist wire, fashioning a loop for hanging in the process.

Hang the chime on a hook, kitchen towel rack, or whatever, and level the tiers by pulling *down* on the rims where they are too high, holding onto the "neck" firmly as you do it.

TRIM WITH A BOW

Cut two 14″ strands of yarn and fold them as shown in Fig. 11. Place them front and back of the wire binding, catch in the two fringed bells, and tie all firmly in place with a 6″ strand of yarn. If the fringed bells tend to look droopy, insert a length of florist wire through them to hold them out perkily.

Time to celebrate!

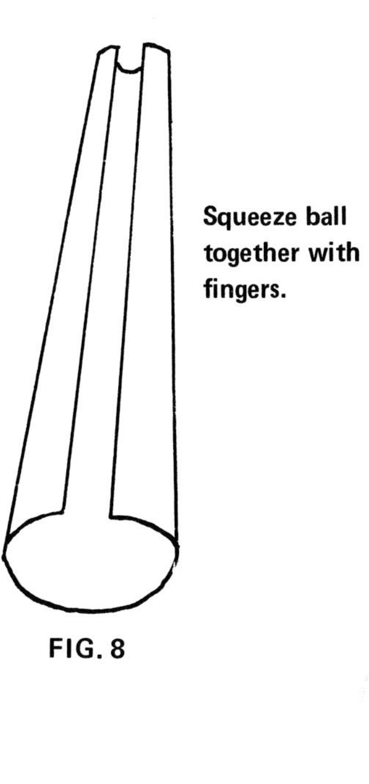

Squeeze ball together with fingers.

FIG. 8

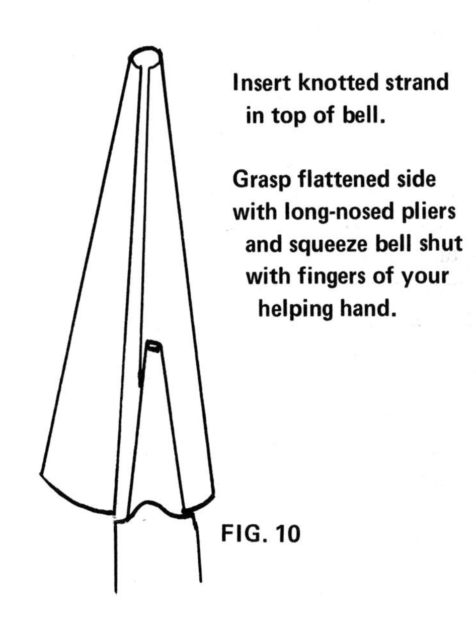

Insert knotted strand in top of bell.

Grasp flattened side with long-nosed pliers and squeeze bell shut with fingers of your helping hand.

FIG. 10

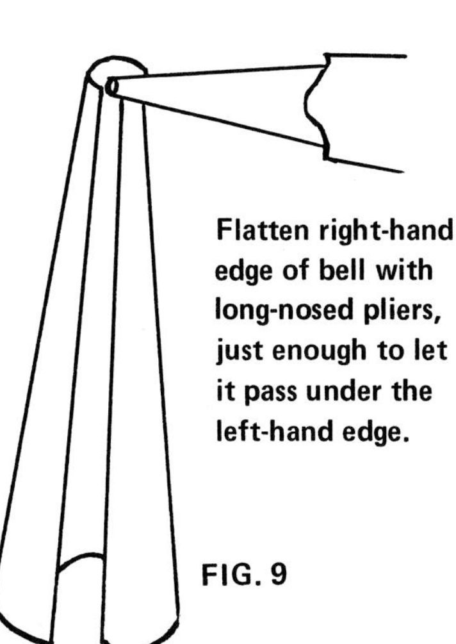

Flatten right-hand edge of bell with long-nosed pliers, just enough to let it pass under the left-hand edge.

FIG. 9

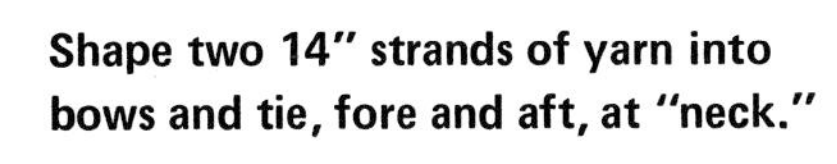

Shape two 14″ strands of yarn into bows and tie, fore and aft, at "neck."

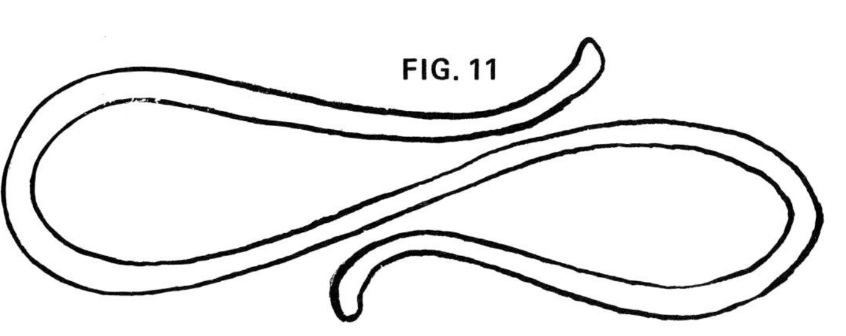

FIG. 11

HYPERBOLE A Figment of the Imagination

Conceived as a birthday present for the television star you see peering out from the petals, this swirling sunflower represents the alpha and omega of tincrafting. Another of Margery Strout's originals, may it be an inspiration to you, as, indeed, I hope the whole book will prove to be.

SOURCES OF SUPPLY

Most tools and materials needed for tincrafting are purchasable at hardware and stationery stores, but the following sources are recommended for your convenience.

Klenk Aviation Snips and other tools:
Welch's Hardware
21 Main Street
Westport, Connecticut 06880

Art supplies, including Architect's Linen, brushes, Talens Transparent Glass Paints, Krylon and Illinois Bronze Spray Paints:
Koenig Art Shop, Inc.
166 Fairfield Avenue
Bridgeport, Connecticut 06611

Cabochon glass stones and Epoxy 220:
Sy Schweitzer & Co., Inc.
P. O. Box 106
Harrison, New York 10528

Jewelry findings, including Nu-Gold wire and maple dapping block:
Allcraft Tool & Supply Co.
20 West 48th Street
New York, New York 10036

Book Sources:
Noah and His Ark by Reinhard Herrmann
The Scribner Book Store
597 Fifth Avenue
New York, New York 10017

The Art of Tray Painting by Maria D. Murray
Keeler's Paint Works, Inc.
40 Green Street
New London, Connecticut 06320

index